Philosophy and Biodiversity

This important collection focuses on the nature and importance of biodiversity. The concept is clarified and its intrinsic and instrumental value is discussed. Even though the term biodiversity was invented in the 1980s to promote the cause of species conservation, discussions on biological diversity go back to Plato.

There are many controversies surrounding biodiversity. Several of them are examined here: What is worthy of protection or restoration and what is the acceptable level of costs? Is it permissible to kill sentient animals to promote native populations? Can species be reintroduced if they disappeared a long time ago? How should the responsibilities for biodiversity be shared?

The book will be of interest to philosophers of science and biologists, but also to anyone interested in conservation and the environment.

Markku Oksanen is Senior Lecturer in Philosophy at the University of Kuopio, Finland.

Juhani Pietarinen is Professor Emeritus of Practical Philosophy at the University of Turku, Finland.

T0273852

CAMBRIDGE STUDIES IN PHILOSOPHY AND BIOLOGY

General Editor
Michael Ruse   *Florida State University*

Advisory Board
Michael Donoghue   *Yale University*
Jean Gayon   *University of Paris*
Jonathan Hodge   *University of Leeds*
Jane Maienschein   *Arizona State University*
Jesus Mosterin   *Instituto de Filosofia (Spanish Research Council)*
Elliott Sober   *University of Wisconsin*

Alfred I. Tauber   *The Immune Self: Theory or Metaphor?*
Elliott Sober   *From a Biological Point of View*
Robert Brandon   *Concepts and Methods in Evolutionary Biology*
Peter Godfrey-Smith   *Complexity and the Function of Mind in Nature*
William A. Rottschaefer   *The Biology and Psychology of Moral Agency*
Sahotra Sarkar   *Genetics and Reductionism*
Jean Gayon   *Darwinism's Struggle for Survival*
Jane Maienschein and Michael Ruse (eds.)   *Biology and the Foundation of Ethics*
Jack Wilson   *Biological Individuality*
Richard Creath and Jane Maienschein (eds.)   *Biology and Epistemology*
Alexander Rosenberg   *Darwinism in Philosophy, Social Science, and Policy*
Peter Beurton, Raphael Falk, and Hans-Jörg Rheinberger (eds.)   *The Concept of the Gene in Development and Evolution*
David Hull   *Science and Selection*
James G. Lennox   *Aristotle's Philosophy of Biology*
Marc Ereshefsky   *The Poverty of the Linnaean Hierarchy*
Kim Sterelny   *The Evolution of Agency and Other Essays*
William S. Cooper   *The Evolution of Reason*
Peter McLaughlin   *What Functions Explain*
Steven Hecht Orzack and Elliott Sober (eds.)   *Adaptationism and Optimality*
Bryan G. Norton   *Searching for Sustainability*
Sandra D. Mitchell   *Biological Complexity and Integrative Pluralism*
Greg Cooper   *The Science of the Struggle for Existence*
Joseph La Porte   *Natural Kinds and Conceptual Change*
Jason Scott Robert   *Embryology, Epigenesis and Evolution*
William F. Harms   *Information and Meaning in Evolutionary Processes*

Philosophy and Biodiversity

Edited by

MARKKU OKSANEN

University of Kuopio

JUHANI PIETARINEN

University of Turku

CAMBRIDGE
UNIVERSITY PRESS

CAMBRIDGE UNIVERSITY PRESS
Cambridge, New York, Melbourne, Madrid, Cape Town, Singapore, São Paulo

Cambridge University Press
The Edinburgh Building, Cambridge CB2 8RU, UK

Published in the United States of America by Cambridge University Press, New York

www.cambridge.org
Information on this title: www.cambridge.org/9780521804301

© Cambridge University Press 2004

This publication is in copyright. Subject to statutory exception
and to the provisions of relevant collective licensing agreements,
no reproduction of any part may take place without the written
permission of Cambridge University Press.

First published 2004
This digitally printed version 2007

A catalogue record for this publication is available from the British Library

Library of Congress Cataloguing in Publication data

Philosophy and biodiversity / edited by Markku Oksanen, Juhani Pietarinen
    p.   cm. – (Cambridge studies in philosophy and biology)
Based on a seminar held in Aug. 1999 at the University of Turku, Finland.
Includes bibliographical references and index.
ISBN 0-521-80430-2
1. Biological diversity – Philosophy. I. Oksanen, Markku, 1966– II. Pietarinen, Juhani.
III. Series.
QH541.15.B56P48 2004
333.95–dc22   2003065423

ISBN 978-0-521-80430-1 hardback
ISBN 978-0-521-03914-7 paperback

Contents

Contents

List of Figures and Tables

List of Contributors

Finn Arler, M.A. and Ph.D. in Philosophy, Associate Professor at Aalborg University, Denmark. He teaches Environmental Ethics and related subjects at several institutions. He is coordinator of education in Human Ecology. From 1997 to 2001 he was associated with the interdisciplinary research project "Boundaries in the Landscape," which is part of the Danish Research Council's program on "Man, Landscape and Biodiversity." He has written numerous articles, mainly in Danish and English, and has edited several books. The latest of these are *Cross-Cultural Protection of Nature and the Environment* (1997) and two Danish anthologies on *Environment and Ethics* (1998) and *Human Ecology. Environment, Technology and Society* (2002). He is currently working on a book on biodiversity and ethics.

Robin Attfield is Professor of Philosophy at Cardiff University, where he has taught philosophy since 1968. His publications include *The Ethics of Environmental Concern, The Ethics of the Global Environment and Environmental Ethics: An Overview for the Twenty-First Century*. He has contributed to a range of philosophical journals including *Ethics, Environmental Ethics, Environmental Values, Inquiry, Journal of Applied Philosophy, Journal of Value Inquiry, Metaphilosophy, Mind, Nous, Philosophy*, and *The Philosophical Quarterly*. Current areas of research interest include the ethics of global warming, meaningful work, the metaphysics of value-talk, environmental ethics, creation and evolution, and Rousseau's deism.

Dieter Birnbacher is Professor of Philosophy at Heinrich Heine University, Düsseldorf, Germany, where his main areas of research and teaching are ethics, applied ethics, and anthropology. He has published a number of books, among them books on Wittgenstein, Schopenhauer, ethics, environmental ethics, medical ethics, and the theory of action. His most recent publication is *Analytische Einführung in die Ethik* (Berlin/New York: de Gruyter, 2003).

Jed Bultitude is Faculty Head of Conservation, Animal Science and Horticulture at Otley College, Suffolk. His research interests include the ecology of herbivory and wildfires in northern boreal forests, Finland. He has published a number of papers on the subject and presented at international conferences on ecology.

Kim Cuddington is an Assistant Professor at Ohio University in the Department of Biological Sciences and Quantitative Biology Institute. Her research interests include spatial population dynamics, scales of environmental variation, and the role of metaphor and models in biology. Her articles have been published in journals such as *American Naturalist, Proceedings of the Royal Society B, Theoretical Population Biology*, and *Biology & Philosophy*.

Christian Gamborg has a Ph.D. in Bioethics and Silviculture from the Royal Veterinary and Agricultural University in Copenhagen and is a research scientist at the Danish Forest and Landscape Research Institute and at the Centre for Bioethics and Risk Assessment. His research interests include ethics, sustainability, and biodiversity in relation to land use, forests, and natural resource management.

Yrjö Haila earned his Ph.D. in ecological zoology at the University of Helsinki in 1983 and has been Professor of Environmental Policy at the University of Tampere since 1995. He has published articles on bird ecology, habitat fragmentation, conservation, social and philosophical dimensions of ecology and environmental issues, and environmental policy. *Humanity and Nature. Ecology, Science and Society* (Yrjö Haila and Richard Levins) was published in 1992 (London: Pluto Press).

Peter R. Hobson is a Lecturer in Conservation Management at Otley College, Suffolk. In addition to contributing as guest lecturer on the international short course on boreal forest biodiversity, Oulu University, Finland, he provides training in habitat assessment for students carrying out international conservation research projects under sponsorship of British Petroleum, Fauna & Flora International, and the Royal Geographical Society. Currently, his research interests are focused on wildfire ecology in boreal forests, Finland, and the historical ecology of ancient woodlands in the United Kingdom and other European countries. He has published a number of papers in these fields and has presented his findings at international conferences.

Julia Koricheva is a Docent in Ecology and an Academy Research Fellow at University of Turku, Finland. Her main research interests are the relationship between biodiversity and ecosystem functioning, plant-herbivore interactions, and meta-analysis and research synthesis. She has contributed to a range of ecological journals including *Ecology, Oecologia, Oikos, Ecology Letters*, and *Evolutionary Ecology* and to the book *Biodiversity and Ecosystem Functioning: Synthesis and Perspectives* (New York: Oxford University Press, 2002).

Keekok Lee is visiting chair in philosophy at the Institute for Environment, Philosophy and Public Policy, University of Lancaster. Interests include environmental philosophy and philosophy of technology, focusing on the ontological distinction between nature and artefacts. Professor Lee's most recent major publication is *Philosophy and Revolutions in Genetics: Deep Science and Deep Technology* (London: Palgrave, 2002).

Gregory M. Mikkelson is an assistant professor with a joint appointment in the Department of Philosophy and the School of Environment at McGill University in Montréal, Québec, Canada. He has an M.S. in Ecology and Evolution and a Ph.D. in the Conceptual Foundations of Science, both from the University of Chicago. Current projects include "objectivity and individualism in environmental ethics" and "biodiversity vs. the automobile."

Markku Oksanen teaches philosophy at the University of Kuopio, Finland. He is a Docent in Environmental Philosophy at the University of Turku, Finland. He has been a researcher in the Finnish Biodiversity Research Programme, based in the Department of Philosophy, University of Turku. His main research interests are environmental philosophy and environmental political theory. His articles have been published in journals such as *Ambio* and *Environmental Values* and in anthologies on environmental ethics and green political theory.

Juhani Pietarinen is Professor (emeritus) of Practical Philosophy, University of Turku, Finland. His research has included work on bioethics, environmental ethics, social philosophy, and history of philosophy. Publications include *Lawlikeness, Analogy, and Inductive Logic* (Amsterdam: North-Holland,

1972); *Perspectives on Human Conduct* (Leiden: E. J. Brill, 1988) edited (with L. Hertzberg); *Genes and Morality: New Essays* (Amsterdam: Rodopi, 1999) edited (with V. Launis and J. Räikkä).

Kate Rawles was a lecturer in environmental philosophy at Lancaster University, United Kingdom, for ten years, specializing in environmental ethics, ethical issues in sustainable development, and animal welfare. She left Lancaster in January 2000 to further pursue these practical aims and now works entirely freelance as a lecturer and consultant. In 2002 she received a major grant from NESTA (National Endowment for Science, Technology and the Arts) to develop "Outdoor Environmental Philosophy."

Michael Ruse is Lucyle T. Werkmeister Professor of Philosophy at Florida State University. He began his teaching career at the University of Guelph, where he taught for thirty years. He is a Fellow of the Royal Society of Canada. He writes on the nature of science, in particular evolutionary biology, and the nature of value. A list of his most recent books includes *Can a Darwinian Be a Christian? The Relation between Science and Religion; Mystery of Mysteries: Is Evolution a Social Construction?; Monad to Man: The Concept of Progress in Evolutionary Biology.*

Peter Sandøe was educated at the University of Copenhagen and at the University of Oxford. From 1992 to 1997 he was Head of the Bioethics Research Group at the University of Copenhagen. He is now Professor in Bioethics at the Royal Veterinary and Agricultural University in Copenhagen and director of the interdisciplinary Centre for Bioethics and Risk Assessment. Since 1992 he has served as Chairman of the Danish Ethical Council for Animals and has been president of the European Society for Agricultural and Food Ethics.

Helena Siipi is a graduate student in philosophy at the University of Turku, Finland. Her research interests include bioethics, environmental philosophy, moral philosophy, and philosophy of action. She is preparing a doctoral thesis on bioethical arguments appealing to naturalness, unnaturalness, and artificiality.

Acknowledgements

This volume is the end result of a seminar entitled "Philosophy and Biodiversity," which the editors organized at the University of Turku, Finland, in August 1999. Many of the contributors to the volume were also participants in the seminar, which was arranged as part of a multidisciplinary Finnish Biodiversity Research Programme, FIBRE (1997–2002). We are grateful to the Academy of Finland for their financial support (project numbers 39456 and 45652). We would also like to thank Hanna-Mari Laine for her help in arranging the meeting and Helena Nissinen for her technical assistance in editing. Dr. Timo Vuorisalo has tirelessly responded to numerous questions from Markku. Michael Ruse, General Editor of the *Cambridge Studies in Philosophy and Biology* series, has been understanding and encouraging. The comments we received from anonymous referees have been extremely valuable in organizing the book and developing the content of individual chapters.

Philosophy and Biodiversity

Biodiversity Considered Philosophically

An Introduction

MARKKU OKSANEN

1. BIODIVERSITY IN THE HUMAN MIND

Biodiversity is peculiar in the sense of being both novel and traditional at the same time. The emergence of the *term* from the discipline of conservation biology is well documented in current history (see Takacs 1996). It is a neologism, dating back to 1985 when Dr. Walter G. Rosen coined it while planning a conference that aimed to bring together what was known about the state of biological diversity on Earth (Wilson 1988, vi). The conference, the National Forum on BioDiversity, was held in Washington, D.C., in September 1986 and the proceedings of this meeting were also titled *Biodiversity* (Wilson 1988). So, biodiversity is a contraction of biological diversity.

A rough *idea* of biological diversity, and thereby of biodiversity, has existed in the human mind ever since evolution endowed our hominid ancestors in the phylogenetic tree with adequate cognitive abilities, in particular that of classification. Therefore, any attempt to definitively date when humans first conceived of nature as diverse is doomed to fail: we live from and within the world of diversity and we are a part of that totality. Some scholars, finding support from different theoretical standpoints such as evolutionary epistemology[1] and cognitive anthropology (Atran 1998), have argued that the human mind has evolved so as to be receptive to nature's diversity and it has a natural capacity, or even "innate dispositions" (Ruse 1989, 189), to organize different elements in nature into handy mental tool-kits that help humans to survive and, ultimately, to live well. Other scholars reject these evolutionary accounts of our cognitive faculties and think of them as acquired and culturally transmitted (e.g., Maffie 1998).

I would like to thank Juhani Pietarinen, Helena Siipi, and Timo Vuorisalo for their helpful comments and Niall Scott for checking the language.

1

Whatever we think of the origin of this ability, both sides must admit that humans need organisms for food, fiber, medicines, tools, and many other purposes. To utilize natural diversity, we have to categorize things; to categorize, we need the criteria of similarity and difference, by means of which we can distinguish edible types from nonedible, useful types from useless, dangerous types from harmless and so on. As Wilson (1994, 40) puts it, "In all cultures, taxonomic classification means survival." Thus, the categorization of the biotic world involves knowledge about it that promotes the evolutionary success of our kind by reducing uncertainties of living. Although the primary motive for categorization might have been practical, it has also served many other purposes, as people distinguish holy or sacred types from profane and beautiful types from ugly, and so on. These categories comprise the cultural dimension of human existence and it is by no means obvious in what way, if any, they are related to the human evolutionary process.

The diversity of life is evident for us at the level of common-sense perception of reality. Thus, biological reality does not consist of unidentifiable objects. Kim Sterelny (1999, 119) has used the term "phenomenological species," as distinct from "evolutionary species," to point out that living organisms have such salient properties that for us the living world contains "identifiable clusters of organisms." The allure of categorization is so strong that even when we are willing to emphasize the individuality of living beings, we tend to delineate them in terms of natural kinds or sortals and identify them as members of certain categories; the idea of "bare particular" is doubtful (see J. Wilson 1999, 16–21). To know an entity is to know it according to its general properties that are denoted by generic terms, such as species membership; thus, to know something is to be informed enough to classify it. When we do not know an entity's kind, we are want of the crucial piece of information to entertain it and we cannot conceptualize it: it remains as a strange or mysterious object. The notion of biodiversity, particularly in folk biology, is a mid-level concept that applies to organizing the apparent resemblance and difference of things. The notion makes sense only within an apprehension of the world that neither regards each individual component of reality as "bare particular" nor the system of nature as a tightly functioning whole, in which any component, or sets of components, cannot be individuated. In brief, biodiversity, both as a vernacular and a scientific concept, is about the classification of perceptible things and phenomena, especially species.[2]

There are many ways of approaching the concept of diversity. We can say, in general, that the categories of nature's diversity constitute scientific knowledge when established according to the rules and standards of scientific research, and folk-biological knowledge when they are established outside the institution of science. In illiterate communities, folk-biological knowledge is delivered from one generation to another through oral tradition. In literate communities the means of knowledge dissemination are more varied for obvious reasons. Western philosophers and scientists have not, in general, acknowledged these "common-sense" achievements, but during the last decade or so, some signs of change have become obvious, and the concept of biodiversity has played a vital role in the course of change. On the one hand, this change goes in tandem with a growing commercial interest in nature's biotic treasures and their potential industrial use; through millennia traditional peoples have acquired basic knowledge of the pharmaceutical, cosmetic, and agricultural uses of various species and varieties of plant and animal life (see, e.g., Swanson 1995; Brush and Stabinsky 1996). On the other hand, it has been noticed that the folk-biological classifications were at times done so well that they coincided with the scientific knowledge (see Medin and Atran 1999), although the reasons for classifying may have fallen short of the standards of biological science.

Views on the philosophy of biology are also changing and such stances as pluralistic realism or "promiscuous realism" have gained support (Dupré 1981, 82). To put it simply, these positions are in favor of the claim that there are many different but defensible ways of classifying nature's diversity. This may imply a certain degree of tolerance and greater understanding of folk-biological classification (cf. Dupré 2002; R. A. Wilson 1999). On the negative side of folk-biological classification, its evaluative dimension is highly selective and typically it manifests many other values and beliefs that are susceptible both from the scientific and conservation points of view. But scientific classification also serves many utilitarian purposes. There are traces of both of these tendencies in the early modern age of botany (Tudge 2002, 21) and they are discussed in the article on Rousseau, for example (see Arler, this volume). Scientific classification rests, however, on a very peculiar idea, that of fully stretched self-criticism, according to which the apparent similarity between living beings can turn out to be illusory and virtually all systems of classification are fallible: one day there is only one species of the African elephant, the next day researchers distinguish between two species of the African elephant, between the forest and the savannah elephant (Roca et al. 2001), that are morphologically distinct and occupy different ecological niches. The replacement of an old belief by a new one because of the discovery of these essential differences is usually interpreted as scientific progress. Despite the fact that the idea of scientific progress and systematic scrutiny can be incongruent with the most conservative systems of folk biology, there

3

is no point in deeming folk-biological systems of belief as constant. Both scientific and folk-biological systems of belief are more or less flexible.

All in all, irrespective of the validity of folk-biological classification, the mere existence of it confirms the idea that the interest of ancient philosophers and naturalists in nature's diversity did not appear out of the blue and the current scientific interest in biodiversity can be seen at the other end of the continuum.[3]

Nature was the predominant concept in classical Greek philosophy from the very beginning. The pre-Socratic philosophers, for instance, assumed that they could identify some primitive element, or elements, of which the world was built. The speculative metaphysical investigation of nature evolved into natural history and into the science of biology and ecology by the nineteenth century. It is telling that in 2001, just fifteen years after the invention of the term biodiversity, a five-volume *Encyclopedia of Biodiversity* was published. Moreover, thousands of scientific articles, as counted by Julia Koricheva and Helena Siipi in their contribution "The Phenomenon of Biodiversity," have been published. Some of these have been published in newly established journals that include "biodiversity" in their titles. Other large-scale projects are on their way to being accomplished, such as the enlargement of the above-mentioned *Encyclopedia of Biodiversity* to an electronic version and the enterprise to make an inventory of all species on Earth.[4] As I see it, without the long preceding history and the established tradition of natural history, broadly understood, nothing like this may have happened, at least not so quickly. Biodiversity has become a buzzword, that is, a currently fashionable expression or a catchword. As is the case with buzzwords generally, biodiversity has also been given innumerable definitions, some of which have grown out of the original context, decreasing its usability. In the opening chapter Koricheva and Siipi provide a survey of this use of the focal concept and analyze how the meaning given to it implies variation in conservation policies.

By coining the new concept, the conservation biologists had a mission in mind: to promote the cause of conservation and to alarm the decision makers about the biological diversity "crisis," as E. O. Wilson (1985) and many others have labeled it (see Haila, in this volume). Thus, biodiversity, the neologism, is a value-laden notion that manifests both the sense of wonder before diversity and the worry over its loss. It was the rapid, mainly anthropogenic, decline of biological diversity that induced the U.S. scientists to invent the catchword and to launch a campaign. What followed can easily be deemed an academic success story, irrespective of the unfortunate background of this enterprise, as it led to worldwide concerted action to block the declining trend. The most notable attainment thus far is the *Convention on Biological Diversity*

4

(CBD) that was signed at the *United Nations Conference on Environment and Development*, held in Rio de Janeiro in 1992. The CBD has three main objectives: to conserve biodiversity, to enhance the sustainable use of its components, and to share the benefits arising out of the use of genetic resources fairly and equitably (see Glowka et al. 1996).

The CBD begins with the definition of biodiversity that has been widely used. Article 2 defines biological diversity as "the variability among living organisms from all sources including, *inter alia*, terrestrial, marine and other aquatic ecosystems and the ecological complexes of which they are part; this includes diversity within species, between species and of ecosystems." Most contributors to this volume and biodiversity textbooks in general tend to follow this definition or simplify it, for instance, as follows: biodiversity refers to the whole variety of life on Earth and to its physical conditions (cf. Perlman and Adelson 1997). By the end of 2002 the convention has been approved by 186 nations. (Ironically, the most notable opposition to it came at that time from the country in which the concept originated, as the then-president George Bush refused to sign it, see Porter, Brown, and Chasek 2000, 124–30.)

Because the history of the neologism "biodiversity" is short and well known, it has often been used as a case of the social construction of environmental problems. This is the starting point of Yrjö Haila's chapter, "Making the Biodiversity Crisis Tractable: A Process Perspective." Drawing from studies on science, Haila claims that biodiversity is above all a dynamic social construction that has become "the organizing center" of various environmental and social concerns. He examines "how the process of construction has fed back to the understanding of the issue itself." He goes on to scrutinize more substantial ecological issues and criticizes the views of prominent conservation biologists, in particular E. O. Wilson and Paul Ehrlich. Like Koricheva and Siipi, Haila offers a comprehensive treatment of the biodiversity problematique that takes a stance on many singular issues. He ends by emphasizing the process nature of biodiversity and the importance of critical discussion regarding the objectives of conservation by means of which weaknesses in opposing extremist stances would become more clear.

2. BIODIVERSITY IN A WORLD OF CHANGE AND CONSTANCY

Given the history outlined above, Sarkar's (2002, 132) remark that "Biodiversity must be analyzed in the context of conservation biology" becomes incontestable. What, then, is philosophically fascinating about biodiversity that goes beyond the burning practical concerns of conservation biology? I think that simply the existence of this volume offers a better answer than I could ever provide here, but let me think about it for a moment. This motivating question is in the background of other questions that I will introduce in the remainder of this chapter.

To begin with, if "the task of conservation is to conserve biodiversity" (Sarkar 2002, 133), it raises the question of what exactly is to be conserved. Ideally we would have a precise operative, hierarchical formulation of what "biodiversity" comprises. The vastness of the extension of the concept biodiversity undermines this prospect and brings in convention: we have to make choices. Although we have the global convention on biodiversity, it is less likely that we will have universally shared biodiversity preservation policies that even include conservation priorities in trade-off situations. Therefore, any answer to the question "What is biodiversity?" has an evaluative dimension (see chapters by Koricheva and Siipi, Haila, Hobson and Bultitude, Rawles, Gamborg and Sandøe, in this volume).

Focusing merely on species, any such attempt to provide an operative definition first leads to systematics, the objective of which is to study and classify the earth's living beings. How, then, does one distinguish between different kinds of organisms? Do natural kinds have essences that are typical of them and only of them? Presuming the traditional realist position according to which species are natural kinds that exist independent of our perception and beliefs, on the one hand, how then does one identify categories that correspond with reality? If we presume, on the other hand, that species are human constructions, it gives rise to many other questions: Is there any truth-value in taxonomic statements? If not, are we then allowed to classify entities however we like? Or should we be paying attention to either individuals or populations in the first place? These questions have been continuously tackled by both taxonomists and philosophers of biology (see, e.g., R. A. Wilson 1999), and answers to them form different background assumptions in conservation biology.

Things get more complicated when the notion of biodiversity is not limited to species; what goes below (e.g., genetic diversity) and above (e.g., ecosystem diversity) this basic unit of categorization is also relevant. Biodiversity exists at different levels of organization, that is, in historically varying genetic lines and communities. The issue of genetic diversity coincides with the issue of species diversity to some extent: how to constitute a scientifically purposeful and/or a policy-relevant distinction between evolutionary significant units within the same species (e.g., subspecies). As to the communities, their classification is also troublesome because there are no sharp lines in nature, but nature constitutes a system made up of interconnected, interdependent, and co-evolving units.

The understanding of biodiversity depends a great deal on the perspective we have chosen. Is biodiversity, above all, a global concept, requiring an objective, disentangled perspective, or can biodiversity be understood locally, from within the biogeographic locality? I think that both perspectives are of importance, and to confront the global perspective with the local one is unfruitful (see Attfield in this volume). Rather it should be seen as an interplay between the concrete and the abstract, between actually existing entities and theoretical idealization. The relationship between the local and the global has various dimensions. Because of the scarcity of empirical evidence, our worry about the future of biodiversity is indefinite: we are often devoid of basic biological information and do not know what we are losing or have already lost and how these changes affect ecosystem functioning (Tilman 2000, 209). Therefore, decisions on land use have to be made on uncertain ground, making mistakes common and requiring re-orientation (see Hobson and Bultitude this volume). Some scholars have suggested adaptive management as a solution to this problem (see Norton 2003). It has a policy dimension. Although conservation policies are implemented locally, in the most abstract sense – and also in a politically significant sense – biodiversity conservation is based on an understanding of diversity of life on Earth: diversity characterizes life. And furthermore, it is based on a particular understanding of the natural world – most of all, on that view the theory of evolution provides. Yet, it is the value of a specific biogeographic locality which matters most, and this requires an ethical judgment. A maximal diversity stored in a gene-bank is a somewhat ambivalent idea and applicable only to a situation in which we are about to lose diversity in its historical context. Thus, zoos and gene-banks should have a very limited role in conservation, and the approach to conservation should rather be ecosystem-centered (Norton 1987).

What is so special about biodiversity that we should pay attention to it and work to preserve it? To answer this question, we need to enter the realm known for the past thirty years as environmental ethics. Environmental ethics systematically examines the ethical relationship between humanity and the rest of nature. The history of systematic environmental ethics is not much older than that of the notion of biodiversity itself. This is the case in particular, if we speak of environmental concern in a global sense, that is, in a sense that is not place-bounded and goes beyond the everyday anxieties about eking out a continuous livelihood from the local environment: the global aspect of environmental *concern* is an essential part of environmental ethics and, in particular, of ethics of biodiversity conservation. This kind of concern for species or for the natural world has not been part of the Western mentality, despite the overwhelming interest in nature and its diversity.

The Greek philosophers reflected on such questions as "Why are there so many kinds?", "What is the relation of a kind to its individual representatives?", "Are these kinds arranged in systematic ways?", and "Why is there order in nature?" Some of their research questions are as topical as ever. The twenty-first century biologists have been looking to explain, for instance, why diversity characterizes life, how different species manage to coexist, and how stability and dynamism relate to each other (see, e.g., Sterelny 1999, 119; Tilman 2000; Brooks and McLennan 2002, 8). Of the Greek philosophers, Aristotle is generally recognized as an originator of the science of biology; his views continued to be powerful until the nineteenth century.

This volume pays special attention to Aristotle's "mentor," Plato. According to Arthur Lovejoy's classical work *The Great Chain of Being*, Plato was the first to make extensive use of the idea that the actual world consists of all possible kinds of living beings. The world is a *plenum formarum*, full of all kinds of beings that ever can exist, and it is the better the more kinds it contains. This idea, which Lovejoy called the Principle of Plenitude, has played a very important role in Western philosophy (Lovejoy 1964; Knuuttila 1999). The principle was adopted by Christian theology, which for centuries taught that the omnipotent God has created the world as perfect and hierarchically structured, admitting of no disappearance of its constituents. It implies the idea that the number of species remains fixed because nothing can disappear from the great chain of being, or *scala naturae*: whenever and wherever a local extinction was noticed, it was nothing but a local matter and the missing species must have survived elsewhere (Moore 1999, 109).

In essence this seems to be Plato's view as well. Does it mean that he thereby was bound to a static conception of nature? Juhani Pietarinen argues in his "Plato on Diversity and Stability in Nature" that Plato's explanation of natural changes has interesting similarities with modern ecological theories. In particular, what Plato calls sensible nature is not "a static collection of various kinds of species," but rather a dynamic system being in a state of constant change but endowed also with a certain ability to resist changes. This kind of "dynamic stability" is essentially dependent on diversity in Plato, according to Pietarinen's interpretation. The relationship between diversity, stability, and dynamism has been the despair of modern ecologists.

It is now unanimously accepted that an extinction of a species can occur. The fossil record clearly speaks for the existence of species that died out long before we entered the scene. Paleontologists have identified five major waves of mass extinction (Sepkoski 1993; see also Boulter 2002, 23–55) and we are, as conservation biologists assert, in the midst of the sixth wave, which is inflicted by humans (Boulter 2002, 189). David Raup (1991, 3–4) has calculated that 99.9 percent of species that ever existed have disappeared. Raup's assumption is of course unproven, as we do not even know the number of currently existing species, so we could not know that of the past (Boulter 2002, 138ff.). The destiny of dinosaurs and mammoths and more recently of dodos, passenger pigeons, and Tasmanian tigers is, nevertheless, familiar to everyone. Even Christian creationists must have reconciled their creeds with the apparent historicity of species. (This is, of course, nothing but an assumption because logically it does not prevent religious zealots from creating new imaginary tales on extinct species in support of their creeds.) Whoever is suspicious about biodiversity conservation policies is motivated by other reasons, such as a belief that it does not pay to conserve, or that humanity can do quite well in a less diverse world. I will return to this issue later on.

Despite the prima facie similarity of the questions being asked by ancient and modern scholars of nature, there are numerous differences in their approaches and answers. For modern ecologists, the meaning of "Why" questions is quite different from that of traditional theologians. The latter have understood them as predominantly metaphysical questions, for example, calling for the underlying plan of the Creator and the idea of cosmic teleology in which each type of being has its own purpose in the functioning of the system, whereas modern scientists reject such ideas and explain the emergence and survival of species with reference to suitable conditions of existence, both biotic and abiotic, that influence the fitness of individuals. This change owes a great deal to Charles Darwin who, however, did not question all prior beliefs.

Although Darwin's theory of evolution and natural selection questioned the traditional theological view, Kim Cuddington and Michael Ruse argue in their chapter "Biodiversity, Darwin, and the Fossil Record" that Darwin still held to the traditional idea of equilibrium. He assumed that the number of species remains somewhat constant even though individual species appear and disappear. When a species is lost, mainly due to competition, it becomes replaced by a new one, often by a near but improved relative of the lost species. An exception to the rule is the case where physical conditions are suitable for species multiplication, for instance through the increases in resource level (in particular, energy). In sum, Darwin was wavering somewhere in between the two opposite poles – the one in which the number of species is eternally fixed (the traditional belief) and the other in which the number of species can increase without any limit (the evolutionary belief). However, his vision of the extant diversity emphasized constancy in the living world: "Darwin paints a picture of nature as essentially stable and predictable in his time, whatever it has been like in the past," Cuddington and Ruse write. The two authors claim that Darwin did not postulate this kind of hypothesis of "dynamic equilibrium" on any evidence, but was rather affected by the pre-scientific, historical views. Cuddington and Ruse also argue that the dynamic equilibrium hypothesis has not lost all of its attraction, as some modern paleontologists, Jack Sepkoski in particular, have "restated" Darwin's position. However, for a modern evolutionary ecologist the idea that the global species number is fixed is absurd: there is no top limit to biodiversity.

Today we look at the phenomenon of diversity through the lenses of evolutionary theory and ecological science. From this perspective biodiversity refers to a set of entities and processes that comprise a complex dynamic system; for this reason it is difficult to define biodiversity in a precise manner, as many contributors to this volume remark. It is an undeniable fact that the diversity in ecological systems is historically varying: when a species is wiped out, there is no necessary substitute for it in the form of new evolving species but, rather, in the form of an invader. Moreover, due to interdependencies between populations of different species, the extinction of one species may imply the same to those that are dependent on it, unless others are able to adapt themselves to the new situation or an immigrant species fills the vacant niche. The diversity of nature varies temporally, yet the understanding of mechanisms of this variation is wanting. Ecologists aim to determine, for instance, the spatial distribution of diversity and relate their findings to various environmental and historical factors so as to explain the origin and the persistence of the extant diversity in a given area of nature. Those ecologists who emphasize competition as the main limiting factor of biodiversity tend to ignore other factors, such as diversity in itself as a "raw material" of further speciation, and this leads them to place emphasis on the notion of stability in the sense of equilibrium.

How does species richness contribute to ecological stability? Many biologists reject the diversity–stability hypothesis, but it has not been outcompeted. To mention just a few recent examples, in a special issue of biodiversity in *Nature*, David Tilman (2000, 208) summarizes the findings of review articles on these experiments: "These reviews show that, on average, greater diversity leads to greater productivity in plant communities, greater nutrient retention in ecosystems and greater ecosystem stability." However, the case may not yet be closed: experimental studies have been conducted for merely a decade and for a relatively limited range of species number. New experiments result in new views. One such experiment shows that there is no positive correlation between great diversity and resilience and resistance to change (Pfisterer and Schmid 2002). The contrasting views reflect differences in a number of questions: whose stability is at issue? A stability of population or of community? What kind of consequences for the ecosystem functioning will result from the loss of biodiversity? What do we mean by stability, exactly?

Gregory Mikkelson discusses some aspects of the debate in this volume ("Biological Diversity, Ecological Stability, and Downward Causation"). His main point is to show the importance of holistic explanations in ecology. For instance, positive or negative effects of diversity in community level on the stability of the component lower-level populations offer an example of downward causation, and the diversity–stability hypothesis an example of holistic explanation. Mikkelson argues that downward causation plays a more important role in nature than scientists have so far recognized, and that neither the same-level causes nor bottom-up causes deserve predominant emphasis in ecology. This means, in effect, rejection of reductionistic explanations. Mikkelson recommends less money for "reductionistically driven ventures" like the Human Genome Project and more for "holistically inspired endeavors, such as what we might call the 'Earth Specionome Project.'"

Fair enough: how could we otherwise learn about the current biodiversity crisis than by aiming to describe the existing species? The project requires an enormous amount of empirical work in the field and taxonomic work in museums before perhaps the greatest question of all for the general public in regard to biodiversity – How much biodiversity does exist on Earth? – is closer to being answered. Yet many scholars are skeptical about the success and rationale of this enterprise and think the question is unanswerable because of its vastness and complexity (Levin 1999, 77). Moreover, all measurements require the identification of units being measured, which necessarily leads to simplifications and thus underestimations (see Purvis and Hector 2000). Haila (in this volume) criticizes the endeavor of naming all the living beings for neglecting the dynamic aspects of the nature of life and being unable to enhance the understanding of those mechanisms that bring about the diversity in systems (also see Hobson and Bultitude's chapter in this volume).

Let us consider this issue from a different angle. In the beginning of this chapter I spoke about the phenomenological species. Common-sense perceptions of the world are often contradictory as there are two strong intuitions that defy each other. When we identify biological entities as members of species, we usually also identify them as individuals or as particulars. To quote Ernst Mayr (1997, 124): "The most impressive aspect of the living world is its diversity. No two individuals in sexually reproducing populations are the same, nor are any two populations, species, or higher taxa. Wherever one looks in nature, one finds uniqueness." If the basic feature of the living world is the *individuality* or *uniqueness* of living beings and of the forma- tions in which they are living, we certainly cannot be concerned with all of them and so the content of the notion of biodiversity remains indeterminate. This is not necessarily a bad thing because the extension of all generic terms is indeterminate; we speak of Finns and Swedes, but any list of Finns or Swedes is incomplete as the scope of reference changes all the time. Still the terms are meaningful. So "biodiversity" is a meaningful term. It is essential to specify what kinds of things we should be concerned about, both morally and scientifically: otherwise the measurement of biodiversity (of an area x at time t) would be reducible to the number of individuals residing in it. To be precise, there would not be diversity of kinds, but of individuals; the more individuals, the more diversity. Biodiversity is not normally understood individualistically, but in terms of categories. However, acknowledging the problems of the species concept and the conjectures about species numbers – the clearest and most commonly used unit to measure biodiversity – are always ambiguous. As many taxonomists tend to emphasize, much of the basic bio- diversity research relies on traditional taxonomic work. More precisely, as long as evolution continues, this work never ends.

3. BIODIVERSITY AT HUMAN MERCY

Humanity is part of a larger system, certain qualities of which are important for human existence and well-being. Stability is clearly such a property because unforeseeable and unpredictable changes in the natural world often result in human suffering: drought, crop failure, decrease in the harvest of fish and so on, often mean that basic human needs cannot be met. In this respect, balance or stability is beneficial to human life. The biosphere, as it exists with its myriad of species, is not stable in the sense of being inflexible. It is a dynamic system with no final end. Hence, the biospheric system survives the tendency of decreasing diversity; biodiversity has declined earlier in prehistoric times, even though probably not at the present rate and speed. This book does not attempt to answer the consequential question: How much of biodiversity and in what time-scale can be lost before we risk our own well-being and survival as a species? I assume that we don't have the precise answer to it, only crude estimations, and moreover, the risk-situation varies spatially and socially (i.e., poorer stratifications of people tend to suffer more than the affluent) (see Haila, Attfield, in this volume). Human knowledge is still very limited, and from this stems one argument for presuming that the best policy is to avoid reducing biodiversity as much as we can. The argument relies on a version of the precautionary principle: lacking precise knowledge, it is rational to assume that we are dependent on rich natural variety. But biodiversity does not only foster humanity, it can be instrumental to human advancement: by losing nature's diversity, many opportunities become perpetually closed for us and for the generations to come (Norton 1999). While biodiversity, to some extent at least, is at human mercy, humans are fully at biodiversity's mercy. The irony in this reasoning is that although humans may have caused the sixth mass extinction, it may lead to new diversification, as has happened earlier in the history of life (Boulter 2002, 82, 191). But this could happen without our species.

The standard arguments for biodiversity preservation are grounded on anthropocentric reasoning, emphasizing the instrumental value of biodiversity. But what kind of value is biodiversity? What kind of elements or units of biodiversity are of intrinsic value, if any, and finally, how do these valuable elements, constituents of total diversity, relate to each other in the action-constraining sense?

The *Convention of Biological Diversity* is ethically an interesting case. While the CBD recognizes noninstrumental value of biodiversity, it is primarily a political-economic statement concerning the allocation of the burdens and benefits of biodiversity policies: the aim is to share them equitably. Before the CBD, genetic resources were open-access resources, over the use of which no one could claim monopoly. According to the CBD, biological resources are subject to national legislation and sovereignty. This was perhaps the main reason for the U.S. resistance to the treaty (see Porter et al. 2000). Moreover, one of the main driving forces of the convention was the vision that mere benevolence is not enough and that sustainable use of units of biodiversity could provide the incentives for conservation. This requires the development of institutions that channel compensation to those parties who act to ensure the future existence of biodiversity (see Swanson 1992; Pearce and Moran 1995). Consider an example. One-half of all pharmaceuticals used today are developed from naturally occurring ingredients, and when a useful organism or a substance is found, it could be worth billions of U.S. dollars/euros. The Rosy Periwinkle that gives us vincristine, a drug used to fight cancer, is a case in point (see Swanson 1995). Traditionally, genetic resources have been public goods; whether their sustainable use requires a change in their institutional status, including introduction of intellectual property rights, is a hot issue at the moment (see Stenson and Gray 1999; Oksanen 2001). Environmental economists in particular have advocated this idea, saying that it would make the conversion of a speciose rainforest to agricultural land, for instance, an unprofitable investment; a better investment would be in local biodiversity, in the form of selling out rights to bioprospecting (such arrangements have taken place, for instance, in Costa Rica [see Takacs 1996] and in Brazil [see Peña-Neira, Dieperink, and Addink 2002]).

There are, of course, innumerable political, legal, economic, cultural, and moral issues concerning the design and enforcement of biodiversity-friendly institutions and practices. The critics of the CBD have pointed out that the value of biodiversity and its units are not reducible to its monetary or use value and that it leads to the commercialization of resources which have thus far been saved from it. Critics have also maintained that discourse on biodiversity has been thoroughly "managerial" (see Rawles, in this volume). The emphasis on instrumental value of biodiversity provides one possible framework within which to read Finn Arler's historical article "Jean-Jacques Rousseau: Philosopher as Botanist," which appears as a chapter in this volume.

Rousseau's eighteenth-century contemporaries were keen amateur naturalists who wanted to exhibit their societal status through their gardens, which were like showrooms for wonders and curiosities of nature. Rousseau despised this attitude; he wanted to promote a disinterested attitude toward natural variety. In his chapter, Arler analyzes Rousseau's key texts on natural history and botany and through them formulates a case for biodiversity preservation, in which encounters with wilderness of the alpine landscape or biodiversity in gardens or in nature could work, in the spirit of emerging Romanticism, as elevating and joyful experiences. As he reads Rousseau, "the external interest in medical virtues and capabilities excluded or impeded any detached view on the plants." If so, their value is dependent on the capability of accomplishing some end. For Rousseau, the ideal motive for botany is "pure curiosity" in which the utilitarian or managerial goals do not distort the study of nature. As Arler emphasizes, however, Rousseau never fully got rid of the anthropocentric attitude toward nature.

In Keekok Lee's chapter, "There Is Biodiversity and Biodiversity: Implications for Environmental Philosophy," the case for noninstrumental value of biodiversity is pushed further. At first glance, the value of biodiversity as a state of affairs is obvious: the more there are diverse kinds of organisms, their habitats and related complex relationships, the better. Nevertheless, it is neither obvious nor conclusively agreed on what to include in biodiversity, and thus what to preserve. In Koricheva and Siipi's chapter, the controversy over the status of human-made biodiversity is provisionally presented. Lee offers detailed reasons why an increase in biodiversity is not always of positive value. Her stance rests on the assumption that a clear distinction can be made between natural and human-made diversity and between nature and culture. Like many other environmental philosophers, she argues that those parts of the natural world are more valuable which are outcomes of nature's own spontaneous forces than parts which are human-generated replicas; we should not substitute a real palm tree with a plastic palm tree because of the inferior value of the latter. This claim is, I believe, universally shared by conservation biologists and environmentalists; however, Lee is more demanding. She makes a distinction between naturally occurring biodiversity and human-made biodiversity. The former consists of various organisms existing independent of human intentions and manipulations, whereas the latter consists of organisms which are human artifacts, that is, they exist as the result of some kind of intentional manipulation (like domestic animals and transgenic organisms). Lee argues that although both kinds of diversity may have instrumental value for humans, only the former has independent moral value, that is, value based entirely on beings whose existence has nothing to do with human intentions. Therefore, all kinds of destruction of such value is morally evil, and, Lee says, appreciating this may provide moral constraints to pause our rush to develop new technologies for transforming nature according to our instrumental values.

Lee's distinction between naturally occurring and human-made entities is one example of contrasting "natural" with "human-generated" biological systems. For Lee, a human-made biological entity is essentially a human artifact, the result of intentional manipulation and control, whereas the term "naturally occurring" refers to entities that exist entirely independent of our intentions, even independent of our existence. In Koricheva and Siipi's chapter, "human-generated" covers more broadly any changes in biological and ecological systems caused by human activities, either intentionally or accidentally, and they call "native" those systems unaffected by such changes. Some other writers speak of "native" individuals and species in contrast to invaders (Cuddington and Ruse; Rawles); native species have a relatively long history in some areas compared to invaders, and the latter (or perhaps both) may or may not have been introduced by humans. Both native and non-native systems may be said to be natural, even when being somehow "human-generated." (Robin Attfield uses the expression "anthropogenic biodiversity" in his chapter.) No attempt is made in this volume to create a unified terminology; the terms are preserved in the form used by the writers.

The distinction between "natural" and "human-generated" usually has a strong normative content. This can be seen, for instance, in the discussion on management of landscapes. For many people, the areas of nature devoid of any visible human "footprint" and in a genuine state of nature are more valuable than areas that are under human management. The policy implication is to promote the quality of wilderness in selected areas of nature whenever possible. The idea of wilderness is, however, deeply problematic in the context that includes millenary tradition of land use. There are many cases in which human practices have increased local biodiversity, and these species and habitats are indeed dependent on the continuation of these practices. In their chapter "Evaluating Biodiversity for Conservation; A Victim of the Traditional Paradigm," two British ecologists Peter R. Hobson and Jed Bultitude consider the nature and rationale of conservation policies regarding landscapes. They analyze both the value dimension of biodiversity conservation in context of selecting sites for conservation, and epistemological dimensions and assumptions of overall conservation aims. Their perspective is characteristically British, implying the emphasis on the recognition of a centuries-long human involvement in the formation of landscapes. This raises a general doubt to the human-nature dualism that other writers support; the critics assert, for instance, that there is simply no human-independent pure nature. This volume's contributors are not unanimous in regard to this issue.

I previously quoted Mayr's view that uniqueness characterizes living entities. In environmental ethics, particularly in biocentric individualism, this biological statement of fact is taken to have moral bearing. In practice, uniqueness implies irreplaceability or nonsubstitutability and thus sets limits upon the treatment of the individual organism: as Paul Taylor (1986) claims, the ethics of respect for nature demands that the individuality of living organisms is recognized. This is in stark contrast to the standard economic understanding of sustainable use of renewable natural resources, stating that the uniqueness of living things does not matter and that any usable thing in nature is substitutable as long as there are individuals, populations, or ecosystems bearing the same valuable properties. These are the extreme positions. Dieter Birnbacher's contribution to the volume, "Limits to Substitutability in Nature Conservation," arises from this constellation.

At first sight, the notion of biodiversity fits for the purposes of sustainable development as it allows, or even encourages, the use of living organisms despite their individuality, but within the limits that they are substitutable by individuals of the same kind or by a kind that perform the same function in an ecosystem. Thus, the value-making properties are being preserved. But is it morally appropriate to consider that organisms are always substitutable? Most environmentalists answer no. One of them, Eric Katz (1997), has called this "the substitution problem." By way of criticizing Katz's rationale of why substitution is a problem, Birnbacher formulates his own vision on the limits and basis of substitution in nature conservation. Particularly in

16

axiological ethics, the value of individuals depends on their relative or nonrelative properties and is not attached to concrete individuals themselves, making properties, not individuals, worthy of protection. If individuals of a certain species are replaceable with other individuals of the species, the species itself may not be replaceable. To not accept this, as the environmentalists tend to think, leads to undermining the value of species if they are considered to be substitutable with a species that fulfills the same function, or niche, in an ecosystem. Birnbacher is critical of standard environmentalist views on substitutability that puts the blame on humans in cases of anthropogenic substitution (when one unit of biodiversity is replaced with another). Birnbacher asks whether the manner of the loss of a unit of biodiversity matters, as the result is always a loss. As he reads Katz, human responsibility covers the forbearances from active destruction of biodiversity and the prevention of natural extinction, the "spontaneous process of substitution," whenever it is within our reach. But, for Birnbacher, Katz would make an ungrounded move, if he rejects this claim (which he ultimately wants to do, otherwise he would not let nature produce a less valuable state of affairs). Birnbacher has a positive attitude toward the idea that humans may actively enhance the evolution of diversity, against Katz and Lee (in this volume) who stress the value of naturalness, provided that it is not at odds with ecological and nonecological values.

Another controversy in determining the value of biodiversity units relates to the issue of to what extent we may use biodiversity classification as a moral guideline in treatment of living entities. Biocentric individualism not only regards living entities as unique, but also as loci of intrinsic value. Taylor (1986) defends the idea of "species impartiality," according to which all intrinsically valuable entities are of equal value and species membership is morally irrelevant property. To adopt this view would cause many ethical problems for biodiversity management in which species membership is one of the main criteria when decisions on the treatment of an organism are made. The animal welfare movement rests on a similar though more focused ideology. It is more focused in the sense that it does not consider that all organisms are of intrinsic value, merely those having sentience, that is, in effect, animals. Peter Singer (1990) and Tom Regan (1988) both espouse egalitarianism, claiming that "All animals are equal." From this perspective Kate Rawles, in her chapter "Biological Diversity and Conservation Policy," criticizes prevailing methods of biodiversity conservation. Many people are confident in their belief that conservation cannot be grounded on the concern for animals and their well-being (a classic statement is Callicott 1980; and its recent reiteration, Rogers 2000, 316). Animal liberationists strictly deny

this claim and argue that the concern for animals is the best available basis
for conservation (Regan), although it sets some limits to potential goals and
methods in conservation. Accordingly, it is Rawles' claim that biodiversity
management methods should not include those methods that lead to killing
of sentient animals: biodiversity conservation does not justify the killing of
animals. Although biocentrists do not oppose the classification of animals
and plants, they are suspicious about its moral appropriateness in ethical
decision making if it involves immoral, as they see it, treatment of sentient
beings. Thus, Rawles takes a position in the environmental ethical debate
among "animal welfarists versus ecocentrists," a debate that may have come
to form the foundation of this recently emerged discipline. She also asserts
that biodiversity conservation is not at the heart of conservation ideology and
disregards biodiversity as the ultimate goal of conservation.

Conservation biology is not only about preventing undesired future scenarios from occurring, but also about realizing, healing, and restoring. The
philosophy and practice of restoration ecology is also historically recent,
dating back to the 1980s. Christian Gamborg and Peter Sandøe's chapter,
"Beavers and Biodiversity: The Ethics of Ecological Restoration," focuses
on one specific issue by means of a case study on the reintroduction of the
beaver in Denmark. Attitudes toward restoration are deeply conflicting and
the understanding of biodiversity differs as well. According to a position,
called the respect-for-nature position by Gamborg and Sandøe, restoration is
suspect because it maintains that anything in nature is substitutable. Whatever
damage or harm has been done to the natural environment can be repaired,
so that the resulting geographic locality is of equal value with the original
site; and whenever this is possible, the altering of ecosystems looks attractive. Thus, the aims of restoration ecology seem to coincide with those of the
Pinchotian, wise-use conservation movement: renewable resources are to be
used for the greatest good of humanity without jeopardizing the possibility
of the posterity to use them too (cf. Norton 1984; Passmore 1980). To put it
simply, as far as the damage can be remedied, nature's goods and services
can be used for our benefit. This is denied by the critics of restoration. They
also claim that the resulting outcome is not as valuable as the original, or it
is a fake (see Elliot 1997; Katz 1997). Although Gamborg and Sandøe recognize many practical and theoretical problems in restoration, their attitude
toward restoration is not disapproving. To overcome this kind of antagonism,
it is necessary that the conflicting notions of biodiversity are fully presented,
paving the way for reconciliation.

Biodiversity is distributed unevenly across the globe. In the beginning of
the twentieth century, the Russian botanist Nikolai Vavilov claimed that most

of the cultivated, edible plants originate in eight geographic areas (Fowler and Mooney 1990, 27–41). According to the modern view, often these Vavilov centers of diversity are also centers of origin, but not always. Conservation biologists have made similar observations recently in their attempt to identify the places of top priority for conservation, that is, places that are particularly rich in endemic species. A group led by Norman Myers found twenty-five such "biodiversity hotspots" that occupy 44 percent of all species of vascular plants and 35 percent of all species in four vertebrate groups. The total hotspot area comprises only 1.4 percent of the land surface of the Earth. Most of these hotspots as well as megadiversity areas are located in third world countries (Myers et al. 2000). Granting the fact that biodiversity is of "a common concern of humankind" and that the responsibility for biodiversity to humanity is shared too, it seems to match up badly with the fact of uneven distribution of biodiversity. The traditional idea in regard to nature conservation is that each country is directly responsible for the diversity of life within its jurisdiction. This seems both a fair division of labor and politically appropriate, considering the existing nation-state system. But, the reverse side of this is the implication that the burden of developing countries is bigger than that of affluent countries, because of the greater diversity of life there. Countries are different – should responsibilities be different too? These considerations create the factual background for Robin Attfield's chapter "Differentiated Responsibilities."

Attfield approaches the issue of responsibility distribution from the perspective of theoretical ethics, in which the divide between universalists and particularists is rehearsed several times. Attfield himself defends "principled universalism." As Attfield says, "theories need to be context-sensitive without abandoning principle in favour of particularity." In the latter part of his chapter, he applies his framework to the Rio Declaration on Environment and Development and finds it to contain reasonable universal principles that are "tailored" to take into account special circumstances of peoples; what their level of ability is; how vulnerable they are; what their position is in terms of power; and so on. Accordingly, although the responsibility for biodiversity is shared, the way it is shared, in effect, varies.

4. BIODIVERSITY IN THIS BOOK

This volume focuses on biodiversity from a philosophical point of view. The notion of biodiversity was invented for purposes of conservation biology, but the concept has grown out of its original domain of use. Today, the research on biodiversity is an interdisciplinary enterprise that involves most academic disciplines from humanities, legal, social, and life sciences. Also this volume is multidisciplinary by virtue of its contributors' academic affiliations.

Biodiversity is a phenomenon that requires a special concern from us, although people disagree on a great number of issues: What does biodiversity mean in practice? What *should* it mean? How much should we be prepared to pay for its protection? How much of research funding should be directed at its study? The list of vital biodiversity-related questions that are worthy of philosophical examination only starts here. Philosophically considered, biodiversity is an important topic of research as it clearly involves many traditional philosophical questions from the existence and nature of species to the rationale of their preservation.

We have divided the book into four parts. Biodiversity is a key concept in conservation biology, and solely for this reason the various aspects and dimensions of the analysis intertwine; consequently, there is some overlap between individual chapters. To say something about the object and objectives of conservation assumes some view of the nature of diversity and its value, and vice versa. As an indication of this, most contributors in this volume have dedicated at least a few lines to discussing the value of biodiversity. The development of the understanding of biodiversity is a historical process and some aspects of it will be elucidated in Part I of this volume (Koricheva and Siipi, Haila). In Part II, both the historical and current understandings of biodiversity will be diagnosed further (Pietarinen, Cuddington and Ruse, Mikkelson). Part III examines the value of biodiversity (Arler, Lee, Hobson and Bultitude, Birnbacher). And finally, Part IV deals with some specific controversies on biodiversity policies (Rawles, Gamborg and Sandøe, Attfield).

NOTES

1. C. A. Hooker (1989, 101) defines evolutionary epistemology as "a doctrine that asserts some interesting relationship between the biological process of evolution and the development of science (and perhaps of cognition more generally)."
2. In the title of a recent book, species are named as "units of biodiversity" (Claridge, Dawah, and Wilson 1997). Ernst Mayr (2002) uses the same expression. See also Wilson (1994, 35–6) and Kareiva and Levin 2003.
3. To procure a general view on the significance of nature's diversity to human existence, the encyclopaedic anthology *Cultural and Spiritual Values of Biodiversity* (Posey 1999) is a good starting point.
4. For this project the Internet is a necessary means of gathering and disseminating information. To see the progress of the project, visit the website: http://www. species2000.org/. The website http://tolweb.org/tree/ is of related interest.

BIBLIOGRAPHY

Atran, S. 1998. Folk Biology and the Anthropology of Science: Cognitive Universals and Cultural Particulars. *Brain and Behavioral Sciences* 21(4):547–69.

Boulter, M. 2002. *Extinction. Evolution and the End of Man.* New York: Columbia University Press.

Brooks, D. R. and D. A. McLennan. 2002. *The Nature of Diversity. An Evolutionary Voyage of Discovery.* Chicago: University of Chicago Press.

Brush, S. B. and D. Stabinsky, eds. 1996. *Valuing Local Knowledge: Indigenous People and Intellectual Property Rights.* Washington, D.C.: Island Press.

Callicott, J. B. 1980. Animal Liberation: A Triangular Affair. *Environmental Ethics* 2:311–28.

Claridge, M. F., H. A. Dawah, and M. R. Wilson, eds. 1997. *Species. The Units of Biodiversity.* London: Chapman and Hall.

Dupré, J. 1981. Natural Kinds and Biological Taxa. *Philosophical Review* XC:66–90.

2002. *Humans and Other Animals.* Oxford: Oxford University Press.

Elliot, R. 1997. *Faking Nature. The Ethics of Environmental Restoration.* London: Routledge.

Encyclopedia of Biodiversity 1–5. 2001. Simon Levin (Editor-in-chief). San Diego, Calif.: Academic Press.

Fowler, C. and P. Mooney. 1990. *Shattering: Food, Politics, and the Loss of Genetic Diversity.* Tucson: University of Arizona Press.

Glowka, L., F. Burhenne-Guilmin, and H. Synge (in collaboration with J. A. McNeely and L. Gündling). 1996. *A Guide to the Convention on Biological Diversity.* Environmental Policy and Law Paper No. 30. Gland, Switzerland: IUCN.

Hooker, C. A. 1989. Evolutionary Epistemology and Naturalist Realism. In *Issues in Evolutionary Epistemology*, K. Hahlweg and C. A. Hooker, eds. Albany: State University of New York Press.

Kareiva, P. and S. A. Levin, eds. 2003. *The Importance of Species: Perspectives on Expendability and Triage.* Princeton, N.J.: Princeton University Press.

Katz, E. 1997. *Nature as Subject: Human Obligation and Natural Community.* Lanham, Md.: Rowman and Littlefield.

Knuuttila, S. 1999. Medieval Theories of Modalities. In *Stanford Encyclopedia of Philosophy.* Website: <http://plato.stanford.edu/entries/modality-medieval/> Read: 20 January, 2003.

Levin, S. A. 1999. *Fragile Dominion. Complexity and the Commons.* Reading, Mass.: Perseus Books.

Lovejoy, A. O. 1964. *The Great Chain of Being. A Study of the History of an Idea.* Cambridge: Harvard University Press.

Maffie, J. 1998. Atran's Evolutionary Psychology: "Say It Ain't Just-So, Joe." *Behavioral and Brain Sciences* 21(4):583–4.

Mayr, E. 1997. *This Is Biology. The Science of the Living World.* Cambridge, Mass.: Harvard University Press.

2002. *What Evolution Is.* London: Phoenix.

Medin, D. L. and S. Atran, eds. 1999. *Folkbiology.* Cambridge, Mass.: MIT Press.

Moore, J. A. 1999. *Science as a Way of Knowing: The Foundations of Modern Biology.* Chicago: University of Chicago Press.

Myers, N., R. A. Mittermaier, C. G. Mittermaier, G. A. B. Da Fonseca, and J. Kent. 2000. Biodiversity Hotspots for Conservation Priorities. *Nature* 403:853–58.

Norton, B. G. 1984. Conservation and Preservation: A Conceptual Rehabilitation. *Environmental Ethics* 8:195–220.

——— 1987. *Why Preserve Natural Variety?* Princeton, N.J.: Princeton University Press.

——— 1999. Ecology and Opportunity: Intergenerational Equity and Sustainable Options. In *Fairness and Futurity*, A. Dobson, ed. Oxford: Oxford University Press.

——— 2003. *Searching for Sustainability. Interdisciplinary Essays in the Philosophy of Conservation Biology.* Cambridge: Cambridge University Press.

Oksanen, M. 2001. Privatising Genetic Resources: Biodiversity Preservation and Intellectual Property Rights. In *Sustaining Liberal Democracy: Ecological Challenges and Opportunities*, J. Barry and M. Wissenburg, eds. London and New York: Palgrave.

Passmore, J. 1980. *Man's Responsibility for Nature.* 2d ed. London: Duckworth.

Pearce, D. and D. Moran. 1995. *The Economic Value of Biodiversity.* London: Earthscan.

Peña-Neira, S., C. Dieperink, and H. Addink. 2002. Equitably Sharing Benefits from the Utilization of Natural Genetic Resources: The Brazilian Interpretation of the Convention on Biological Diversity. *Electronic Journal of Comparative Law* 6.3. Website: <http://www.ejcl.org/63/art63-2.html>.

Perlman, D. L. and G. Adelson. 1997. *Biodiversity: Exploring Values and Priorities in Conservation.* Malden: Blackwell Science.

Pfisterer, A. and B. Schmid. 2002. Diversity-Dependent Production Can Decrease the Stability of Ecosystem Functioning. *Nature* 416:84–6.

Porter, G., J. W. Brown, and P. S. Chasek. 2000. *Global Environmental Politics.* 3rd ed. Boulder, Colo.: Westview Press.

Posey, D. A. (comp. & ed.). 1999. *Cultural and Spiritual Values of Biodiversity. A Complementary Contribution to the Global Biodiversity Assessment.* London: Intermediate Technology Publications; and Nairobi, Kenya: United Nations Environment Programme.

Purvis, A. and A. Hector. 2000. Getting the Measure of Biodiversity. *Nature* 405:212–19.

Raup, D. M. 1991. *Extinction: Bad Genes or Bad Luck.* New York: W. W. Norton.

Regan, T. 1988. *The Case for Animal Rights.* London: Routledge.

Roca, A. L., N. Georgiadis, J. Pecon-Slattery, and S. J. O'Brien. 2001. Genetic Evidence for Two Species of Elephant in Africa. *Science* 293:1473–7.

Rogers, B. 2000. The Nature of Value and the Value of Nature: A Philosophical Overview. *International Affairs* 76(2):315–23.

Ruse, M. 1989. The View from Somewhere. A Critical Defense of Evolutionary Epistemology. In *Issues in Evolutionary Epistemology*, K. Hahlweg and C. A. Hooker, eds. Albany: State University of New York Press.

Sarkar, S. 2002. Defining 'Biodiversity'; Assessing Biodiversity. *The Monist* 85:131–55.

Sepkoski, J. J. 1993. Ten Years in the Library: New Data Confirm Paleontological Patterns. *Paleobiology* 19:43–51.

Singer, P. 1990. *Animal Liberation.* Rev. ed. New York: Avon Books.

Stenson, A. J. and T. S. Gray. 1999. *The Politics of Genetic Resources Control.* Houndsmills: MacMillan.

Sterelny, K. 1999. Species as Ecological Mosaics. *Species. New Interdisciplinary Essays*, R. A. Wilson, ed. Cambridge, Mass.: MIT Press.

Swanson, T. M. 1992. Economics of a Biodiversity Convention. *Ambio* 21:250–7.

Swanson, T. M., ed. 1995. *Intellectual Property Rights and Biodiversity Conservation: An Interdisciplinary Analysis of the Values of Medical Plants*. Cambridge: Cambridge University Press.

Takacs, D. 1996. *The Idea of Biodiversity. Philosophies of Life*. Baltimore, Md.: Johns Hopkins University Press.

Taylor, P. W. 1986. *Respect for Nature. A Theory of Environmental Ethics*. Princeton, N.J.: Princeton University Press.

Tilman, D. 2000. Causes, Consequences and Ethics of Biodiversity. *Nature* 405:208–11.

Tudge, C. 2002. *The Variety of Life. A Survey and a Celebration of All the Creatures that Have Ever Lived*. Oxford: Oxford University Press.

Wilson, E. O. 1985. Biological Diversity Crisis. *Bioscience* 35:700–6.

1988. Editor's Foreword. In *Biodiversity*, E. O. Wilson, ed. Washington, D.C.: National Academy Press.

1994. *The Diversity of Life*. Harmondsworth: Penguin.

Wilson, J. 1999. *Biological Individuality*. Cambridge: Cambridge University Press.

Wilson, R. A., ed. 1999. *Species. New Interdisciplinary Essays*. Cambridge, Mass.: MIT Press.

Part I

Using 'Biodiversity'

1

The Phenomenon of Biodiversity

JULIA KORICHEVA AND HELENA SIIPI

[D]iversity is rather like an optical illusion. The more it is looked at, the less clearly defined it appears to be and viewing it from different angles can lead to different perceptions of what is involved.

Magurran (1988)

The complexity of the biodiversity concept does not only mirror the natural world it supposedly represents; it is that plus the complexity of human interactions with the natural world, the inextricable skein of our values and its value, of our inability to separate our concept of a thing from the thing itself. Don't know what biodiversity is? You can't.

Takacs (1996)

1.1 INTRODUCTION

The neologism "biodiversity" entered the vocabularies of science, bioethics, the media, and the environmentally aware public during the 1980s; its longer version, "biological diversity," has been in use since the late 1970s (Schwarz et al. 1976; States et al. 1978). After the National Forum on BioDiversity in Washington, D.C. in 1986 and the publication of the proceedings of that conference (Wilson 1988), use of the term "biodiversity" increased very rapidly. The number of scientific publications on biodiversity issues has grown exponentially since the late 1980s and currently exceeds 3,000 per year (Fig. 1.1). The term has also quickly gained popularity outside the scientific realm, and has become a household word widely used by politicians, the media, and at least to some extent the general public. Biodiversity has become one of the central issues of scientific and political concern worldwide; it has

We are grateful to Don DeLong, Erkki Haukioja, Stig Larsson, and Irmi Seidl for constructive comments on the earlier versions of the manuscript and to Ellen Valle for checking the English. The authors were supported by the Academy of Finland.

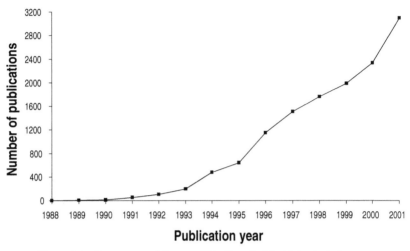

Figure 1.1. Occurrence of the term "biodiversity" in biological abstracts

been acknowledged as a new type of natural resource (Haila and Kouki 1994) and a new target for conservation efforts (Spellerberg 1997, 7).

Despite its wide use, biodiversity still lacks a universally agreed upon definition and is often redefined depending on the context and the author's purpose. For instance, as many as eighty-five different definitions of biodiversity have been reviewed by DeLong (1996). Such great terminological variation is understandable, since concerns for biodiversity relate to several realms of human practice, including conservation, management, economics, and ethics, and thus give rise to different "discourses" (Haila and Kouki 1994). Nevertheless, the lack of a unified fundamental definition, not biased toward any particular discipline, constitutes a serious obstacle to biodiversity research, management, and conservation. The biodiversity crisis cannot be resolved without effective communication and cooperation among different disciplines, organizations, and countries. If biodiversity is defined in fundamentally different ways, agreement on management objectives and strategies for biodiversity conservation may be seriously impaired (DeLong 1996).

The interpretation of the meaning of biodiversity thus represents an important mission for scientists (Bowman 1993; Angermeier 1994) and is the subject of this chapter. We do not intend to provide an exhaustive review of biodiversity definitions; the topic is too broad to be covered in a single chapter, and in any case the question has been addressed in many recent reviews and books (DeLong 1996; Gaston 1996a; Takacs 1996; Swingland 2001). Instead, our aim is to provide a basis for the reconciliation of conflicting views on the

28

biodiversity phenomenon. Our approach to the problem is governed by the belief that the chaos currently reigning in biodiversity terminology is largely due to the confounding on the one hand of the biodiversity definition, operational measures, and biodiversity conservation, and on the other of biodiversity as a scientific concept and as a social/political construct (DeLong 1996; Gaston 1996b). We start with the definition of biodiversity by describing those elements and attributes of the concept that are widely accepted as well as those that are more controversial. We then proceed to the operational measures of biodiversity and discuss values commonly associated with the concept. After that we characterize biodiversity as a social and political construct and discuss ways in which this construct differs from the scientific concept of biodiversity. Finally, we discuss the implications of the biodiversity definition and values for conservation and management.

1.2 DEFINITION OF BIODIVERSITY

The objective of definition is to specify the necessary and sufficient conditions for being the kind of thing the term designates. To define the full scope of the term biodiversity, we shall discuss its components and attributes and compare its meaning to the lexical forms from which the term is derived (the noun stem "diversity" and the prefix "bio-") and to related terms, such as species richness, ecological diversity, biological integrity, and native or natural diversity (cf. DeLong 1996).

1.2.1 Range of Definitions

The various definitions of biodiversity differ from each other primarily in their degree of inclusiveness or concreteness (Table 1.1). The term biodiversity is often interpreted, very broadly, as the richness, variety, and variability of living organisms (Groombridge 1992, xiii; Heywood and Baste 1995, 9; Pearce and Moran 1995, 1; Savage 1995, 673; Jeffries 1997, 3). When used in this way, the term is highly abstract; it refers to all life on Earth and its multiple manifestations in different places and times. In addition, this broad definition acknowledges that the term biodiversity has a built-in reference both to the number of biological components (richness) and to the differences (variety and variability) among them. Less inclusive definitions, however, may restrict the meaning of biodiversity only to the number of biological elements currently present in a certain area (Table 1.1). There is also large variation among definitions with respect to the set of components that are included in the concept of biodiversity. Early definitions tend to be least inclusive and equate

Table 1.1. *Range of Biodiversity Definitions According to Scope and Emphasis on Particular Elements and Attributes*

Aspect	Least Inclusive	→	Most Inclusive
Spatial scale	Local (concrete area)[1]	→	Global (Earth)[2]
Temporal scale	Present diversity[3]	→	Past, present, and future diversity[4]
Components of hierarchy	Species diversity[5]	→	Genetic, taxonomic, and ecological diversity[6]
Number of entities vs. difference between entities	Number only (richness)[7]	→	Number of and degree of difference between entities[8]
Entities vs. processes	Entities only[9] (composition)	→	Both entities and processes[10] (composition, structure, and function)
Biotic vs. abiotic components	Only biotic components[11]		Both biotic and abiotic components[12]
Native vs. human-generated diversity	Only native diversity[13]	→	Both native and human-generated diversity[14]

[1] Schwarz et al. 1976; Eisner (quoted in Takacs 1996, 47); [2] ICBP 1992; [3] Schwarz et al. 1976; [4] Gaston and Spicer 1998; [5] Schwarz et al. 1976; Lovejoy 1980; [6] Harper and Hawksworth 1996; Jeffries 1997; [7] Schwarz et al. 1976; [8] OTA 1987; [9] Angermeier and Karr 1994; [10] Franklin 1988; Noss 1990; [11] DeLong 1996; [12] Noss 1990; [13] Samson and Knopf 1994; [14] Dasmann 1991.

biodiversity with species richness (e.g., Schwarz et al. 1976; Lovejoy 1980). Species, however, represent only one of the many elements of biodiversity, and more recent definitions acknowledge this by including other levels of the biological hierarchy as well (see section 1.2.2).

The attempts by some authors to narrow the scope of the concept of biodiversity either spatially, temporally, or with respect to the number of elements included seem to be driven by the desire to restrict the definition to those aspects which can be assessed empirically (e.g., species richness in a given locality). This approach, however, confounds definition with application, and is partly to blame for the terminological confusion surrounding the concept of biodiversity (DeLong 1996). DeLong (1996) warns that the need to measure biodiversity should not drive the fundamental meaning of the term, and temptations to narrow the scope of the concept for particular management or monitoring purposes should be resisted. We shall discuss operational measures of biodiversity in section 1.3 and practical implications of the definition in section 1.6. There are also more fundamental/conceptual differences in biodiversity definitions, depending on whether they include or exclude processes, abiotic components, and human-generated diversity (Table 1.1). These differences will be discussed in more detail in sections 1.2.3 and 1.2.4.

1.2.2 Hierarchical Structure of Biodiversity

A common feature of most biodiversity definitions is the acknowledgement of the hierarchical structure of its components. For instance, the two most widely cited definitions of biodiversity (by the Office of Technology Assessment 1987 and in the Convention on Biological Diversity, Glowka et al. 1994) describe biodiversity at three fundamental hierarchical levels: genes (within-species diversity), species, and ecosystems. However, a closer look at these three levels reveals that they each have their own hierarchical structure (Heywood and Baste 1995, 9). More recent definitions therefore distinguish three hierarchies: the genetic, the taxonomic or organismal, and that based on ecological diversity (Angermeier and Karr 1994, 691; Heywood and Baste 1995, 10; Jeffries 1997, 74; Table 1.2). These three hierarchies are intimately linked through shared common elements, with the population as the main linking element. The levels within each hierarchy are nested: for instance the phylum comprises classes, which comprise orders; genes comprise nucleotides; the biosphere comprises biomes, and so on. The implications of this hierarchical structure of biodiversity components for biodiversity assessment and conservation are discussed in sections 1.3 and 1.6.

Diversity within each hierarchical level can be characterized in terms of three primary attributes: composition, structure, and function (Franklin 1988; Noss 1990). Compositional diversity as Noss defines it has to do with the identity and variety of elements at each hierarchical level. Structural diversity

Table 1.2. *Hierarchical Structure of Biodiversity Components*

Genetic Diversity	Taxonomic (Organismal) Diversity	Ecological Diversity
Populations	Kingdoms	Biosphere
Individuals	Phyla	Biomes
Genome	Classes	Bioregions
Chromosomes	Orders	Landscapes
Genes	Families	Ecosystems
Nucleotides	Genera	Communities
	Species	Habitats
	Subspecies	Niche
	Populations	Populations
	Individuals	

*Source:*After Angermeier and Karr 1994; Heywood and Baste 1995

refers to the physical organization or pattern of a system, from genetic structure to landscape patterns. Functional diversity involves genetic, ecological, and evolutionary processes, such as gene flow, natural disturbances, and nutrient cycling. This brings up the interesting question of the role of processes in the concept of biodiversity.

1.2.3   Role of Processes

Noss (1990) criticizes early definitions of biodiversity (e.g., the one by OTA 1987) for their failure to mention ecological processes as an important component of the concept of biodiversity. His main argument for the inclusion of processes in the definition of biodiversity is that ecological processes are crucial for the maintenance of diversity of biological elements. However, not all scientists agree that this warrants the inclusion of processes in the definition of biodiversity. For instance, DeLong (1996, 742) argues that "including ecological processes as part of biodiversity because they are important to maintaining biodiversity equates to including potable water and airborne oxygen as components of a human being because they are necessary for the survival of human beings. It confuses definition with functional relationships." Moreover, experimental evidence is beginning to accumulate that communities composed of fewer species, trophic levels, and functional groups demonstrate lower rates of such ecosystem processes as primary productivity, nutrient retention, and decomposition as compared to more diverse communities (reviewed in Schläpfer and Schmid 1999; Tilman 1999). It therefore appears that the diversity of entities at different levels of biological hierarchy determines the rate of ecosystem processes rather than the other way around.

Inclusion of processes in the concept of biodiversity has also been criticized on semantic grounds. For instance, Angermeier and Karr (1994) reason that the term "diversity" is inapplicable to processes because basic evolutionary and ecological processes, such as photosynthesis, speciation, predation, and disturbance, are universal (generic); differences among ecosystems arise due to differences in process rates rather than process occurrence (diversity). Angermeier and Karr therefore suggest that the concept of biodiversity should be limited only to the elements (genes, populations, species etc.), whereas processes should be considered as components of biological integrity. However, as noted by Gaston (1996b), even when biodiversity is defined in terms of entities, it implicitly embraces processes in at least two ways: by recognizing that one dimension of biodiversity is the variety of functions that entities (e.g., species) perform, and through the inclusion of ecosystems as a component of biodiversity, since the concept of ecosystem covers processes as well as entities (Noss 1996).

Another semantic argument against the inclusion of processes in the definition of biodiversity is put forward by DeLong (1996), who reasons that many ecological processes, such as water cycling, soil erosion, wind and fire, are abiotic in origin and thus do not fit the meaning of the prefix "bio-" in the term biodiversity. Using the same argument, DeLong also objects to the inclusion of ecosystems and landscapes in the concept of biodiversity, since these by definition include some abiotic components. He suggests that ecological diversity is a more appropriate term for definitions including ecological processes and ecosystems. Biodiversity should therefore be seen, according to DeLong, as a component of ecological diversity rather than the other way around (cf. Table 1.2).

1.2.4 Native versus Human-Generated Diversity

Deliberate or accidental activities of humans may change the features of biological organisms (as in domestic animals, cultivated plants, human-created hybrids, chimera species) or cause these organisms to move into new surroundings (alien or introduced species). Such human-generated enhancement of diversity may occur at different hierarchical levels, including genetic (gene transfer), taxonomic (alien species), and ecological (landscape modification) diversity. The attitude of philosophers and biologists toward the place of human-generated components in the concept of biodiversity varies from one extreme to the other. Some argue that the concept of biodiversity should be restricted to native biodiversity only (Angermeier 1994; Angermeier and Karr 1994), since "artificial diversity cannot provide the full array of societal values that native diversity does" (Angermeier 1994, 602); moreover, human-generated diversity may represent a threat to native diversity (for instance introduced species may drive native ones to extinction through competition, predation, transmission of diseases and parasites, and/or alteration of the composition, structure, and ecological processes which characterize ecosystems). Others maintain that human-generated elements are not only an important part of biodiversity: "no one can question that diversity exists in each of these cases and that the diversity is biological" (Perlman and Adelson 1997, 60), but also an undervalued one. Concern is often expressed, for instance, that the development of large modern farms leads to dramatic losses of varieties (and thus genetic diversity) of cultivated plants (Myers 1997; Perlman and Adelson 1997, 58–60), and the possible detrimental consequences of these losses for breeding for productivity and resistance to diseases and pests are discussed. Our tendencies to include or exclude human-generated elements from the concept of biodiversity thus seem to mirror significantly the values we assign to different elements (see also section 1.6.2).

The debate over the role of human-generated elements in the concept of biodiversity is also closely related to the question of the place of humans in nature. The answer given distinguishes two schools of philosophy: compositionalism and functionalism (Callicott et al. 1999). In the functionalist's view, humans are a part of nature and "only fellow-voyagers with other creatures in the odyssey of evolution" (Leopold 1970, 116–17; Callicott et al. 1999, 24). According to this view, everything humans do is part of nature, and natural in that primary sense. When we influence other species, this is simply an example of one species exerting a selection pressure on another (Sober 1988, 180). Functionalists therefore tend to consider alien species, domestic species, and even transgenic organisms as part of biodiversity. Compositionalists, in contrast, view humans as a case apart from nature. Albeit an evolved species, *Homo sapiens* is considered to be a destructive force external to the biota because it employs unique, culturally created means of altering the natural environment to suit itself (Callicott et al. 1999, 24). From this point of view all human modifications of nature are artificial, and human-generated elements should not be considered as part of biodiversity. Even if one accepts this dichotomy between the natural and the artificial, however, there still remains the problem of where to draw the line between the two, since humans have directly or indirectly affected almost all ecosystems and species living in them.

As a compromise between these two views, DeLong (1996) has suggested that both native and human-generated diversity should be included in the basic definition of biodiversity, but that a more restrictive term – native biodiversity – be used when discussing biodiversity native to an area. The rest of this chapter, as well as most of the other chapters in this book, deal mainly with native biodiversity.

1.3   OPERATIONAL MEASURES OF BIODIVERSITY

Defined in the previous sections as the richness, variety, and variability of life, expressed across a range of hierarchical scales, biodiversity represents a rather abstract concept. On the other hand, a quantitative assessment of biodiversity is clearly required for purposes of management and conservation, for the evaluation of human impacts on biological systems, and for testing hypothesized relationships between biodiversity and ecosystem functioning (e.g., stability or resistance to invasions). The need to develop operational measures of biodiversity has therefore been repeatedly emphasized (Angermeier 1994; DeLong 1996; Harper and Hawksworth 1996). It is important, however, to distinguish between two ideas related to biodiversity measures: that biodiversity per se can be quantified, and that different elements or attributes of biodiversity can be quantified (Gaston 1996b). The multidimensionality of biodiversity implies that it cannot be reduced to a single figure which would capture all aspects of the concept. No single measure can embrace diversity at all levels of the biological hierarchy; thus, several different measures have to be employed to describe and compare the biodiversity of various areas. In the following discussion we use the expression "biodiversity measures" in this latter, more restricted sense.

There are basically three types of biodiversity measures: those which assess the number of entities at a particular hierarchical level, those which assess the degree of difference between these entities, and those which attempt to combine both aspects. These measures can be further classified according to the level and type of biological hierarchy they assess – genetic diversity, organismal diversity, or ecological diversity (Table 1.3).

The major problem with measures that are based on the number of entities existing at a certain hierarchical level (cardinal measures as defined by Cousins 1991) is that they treat each entity as equal or as contributing equally to the biodiversity of an area. However, this assumption is clearly questionable; at the species level, for instance, the relative contribution of individual species to biodiversity depends on their abundance, spatial and temporal distribution, their phylogenetic distinctiveness (Wheeler 1995, 24; Perlman and Adelson 1997, 27–8), their evolutionary potential (Erwin 1991), and the functional role they perform in ecosystems (Lawton 1994). Similar differences also exist between other elements of the biological hierarchy (the frequencies of a certain gene, for instance, may differ both within and between populations), although less effort has been made to quantify differences between various ecosystems, landscapes, and so on. Ordinal measures (according to Cousins 1991), which incorporate both the number of entities and differences between them, therefore are clearly more meaningful, even though they may incorporate only one aspect of difference at a time (most commonly differences between entities in terms of their abundances), and although few such measures have been developed for hierarchical levels other than genes and species (Table 1.3).

It is clear from Table 1.3 that most operational measures have been developed to assess genetic and organismal/taxonomic diversity. This is not surprising, given that the components of genetic diversity – such as genes,

Table 1.3. *Examples of Biodiversity Measures Classified by Hierarchical Level and Aspect Assessed by the Measure*

Aspect	Genetic Diversity	Taxonomic/ Organismal Diversity	Ecological Diversity
Number of entities	Number of alleles at a given locus[1] Number of genetically distinct populations per area[2]	Species richness[3] Generic diversity[4] Family richness[5]	Habitat diversity[6] Landscape diversity
Measures of difference in:			
– abundance	Frequency of the allele in a population[1]	Evenness (E)[6]	
– spatial distribution	Variance of allele frequencies between populations[1]	Beta diversity[7]	
– phylogeny		Taxonomic distinctiveness[8]	
– functional aspects		Per capita interaction strength[9]	Niche overlap[11]
Combined measures	Heterozygosity at a single locus[1]	Shannon index (H′)[6] Simpson index (D)[6] Hierarchical diversity[10]	

[1] Mallet 1996; [2] Hughes et al. 1997; [3] Gaston 1996c; [4] Prance 1996; [5] Williams et al. 1994; [6] Magurran 1988; [7] Whittaker 1977; [8] Vane-Wright et al. 1991; [9] Paine 1992; [10] Pielou 1975; [11] Hurlbert 1978.

chromosomes, and nucleotides – are "real"; thus, they can be identified, counted, measured. Distinguishing between species and classifying them into genera and families within the hierarchy of taxonomic diversity is already a more difficult task, while the elements of ecological diversity represent convenient abstractions rather than real entities and are thus an even more difficult object for counting and comparison.

It is also not very surprising that the number of species (species richness) remains the most frequently applied measure of biodiversity. The meaning of species richness is widely understood, so there is no need to derive complex indices to express it. It is a measurable parameter, and much information on species richness is already available. In addition, there is evidence that differences in species richness often (although not always) parallel differences in other measures of biodiversity, such as phylogenetic disparity, functional diversity, and diversity of landscape (Gaston 1996c). A major conceptual problem with the use of species richness as a measure of biodiversity is the lack of universal agreement as to the definition of species. Different species concepts are employed for different groups of organisms, making species a poor choice as a currency for comparisons of biodiversity among different biotas (which presumably differ in their proportion of different taxonomic groups). In addition, there are practical problems associated with the use of species richness as a measure of biodiversity. First of all, in species-rich communities complete species counts are often impractical or too expensive. Second, in some groups of organisms (e.g., insects) or types of habitats (e.g., tropical and aquatic) only a small proportion of species has been described. Finally, species richness counts are scale-dependent, in the sense that increases either in the spatial scale of the study or in its duration/sampling effort result in higher species richness estimates.

One possible solution to these problems is to use indirect, surrogate methods to assess species richness. Gaston (1996c) distinguished between three main types of such approaches: methods based on environmental variables, on indicator groups, and on the numbers of higher taxa. All three surrogacy methods, however, share certain limitations (Gaston 1996c). In particular, the underlying relationships between the species richness of a target group on the one hand, and on the other hand environmental variables, the species richness of the indicator group, or the number of higher taxa, appear to be case-specific and work only for certain spatial scales, regions, and taxa. Extrapolation from the relationship beyond these limits may reduce the accuracy of the method.

It should be clear from the previous discussion that there is no single "ideal" measure of biodiversity. Rather, there are many different indices, each with its own advantages and disadvantages. The choice of a biodiversity measure depends on the purpose for which it is to be used (Magurran 1988; Gaston 1996b). Since each biodiversity measure captures only some elements or attributes of biodiversity, by choosing a biodiversity index one is also assigning a value to the particular aspect of biodiversity expressed by this index (Williams and Humphries 1996). Therefore, "both what you are measuring and how you are measuring it reveals something about what you most value" (Gaston and Spicer 1998; see also sections 1.4 and 1.6).

The fact that no single measure can capture all the components and attributes of the concept of biodiversity should not be seen as a weakness but as a manifestation of the concept's flexibility. Since our values and conservation priorities have a tendency to change with time (as evidenced by the gradual abandonment of the strategy of preserving endangered species in favor of the preservation of genetic diversity, ecosystems, and processes), new indices could be developed to assess biodiversity in the light of new conservation strategies (for instance indices of phylogenetic distinctness and functional differences).

1.4 VALUES OF BIODIVERSITY

The term biodiversity carries with it an "inextricable skein of our values and its value" (Takacs 1996; Fig. 1.2). The definition of biodiversity as the variety of life at different levels of the biological hierarchy seems neutral with regard to the values assigned to the different elements. However, our views concerning the role particular elements play in the concept of biodiversity (for instance native vs. human-generated elements) are often affected by the values we place on these elements. This seems to be due to the rather general consensus that biodiversity is a good thing and that its loss is bad (Gaston 1996b). In addition, our choice of operational measures is strongly affected by our values, and scientists often use value arguments in articulating the importance of biodiversity preservation to the public. In the following, we therefore discuss the types of values that are usually associated with phenomena of biodiversity.

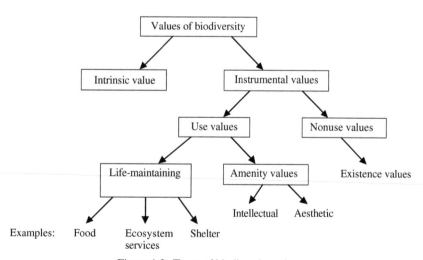

Figure 1.2. Types of biodiversity values

In speaking of biodiversity values, there are two aspects which need to be distinguished: values of individual elements (i.e., a species or an ecosystem), which are the "building blocks" of biodiversity, and values of biodiversity as a whole. We shall maintain this distinction clearly in the following discussion.

1.4.1 Instrumental Values

Something has instrumental value if it enhances the well-being or quality of some valuable entity or state of affairs. Utilitarian value is a species of instrumental value which characterizes objects or states of affairs that are capable of contributing to human well-being (Cox 1997, 174). There is no question that many elements of biodiversity facilitate human welfare or happiness and thus have utilitarian value. Utilitarian values may be so-called use values. We use natural entities as raw materials for food, shelter, clothes, but also for aesthetic pleasure and spiritual inspiration. Sometimes one can even derive pleasure simply from the knowledge that some element of biodiversity, usually a species, exists (Norton 1987, 41–2). This so-called existence value is a nonuse value (Fig. 1.2).

It has also been argued that biodiversity per se, the richness, variety, and variability of biological elements, is instrumentally valuable to human beings (i.e., has utilitarian value). A diverse world provides multiple options for the satisfaction of our needs. The higher the number and the degree of difference between biological elements, the more chances we have to find new biological resources. As the entities through which ecological processes are realized, the elements of biodiversity perform a multitude of functions: for instance they create soil and maintain its fertility, control the global climate, maintain the balance of atmospheric gases, process organic wastes, pollinate crops and flowers, recycle nutrients (Myers 1996). It has been suggested that diversity increases the stability of ecosystems, and thus the stability of these ecosystem services (the diversity–stability hypothesis, McCann 2000; Tilman 2000, 208). The greater the diversity, the more likely it becomes that communities will contain species or functional groups that are capable of differential responses to environmental conditions. Biodiversity may therefore be seen as a form of insurance for proper ecosystem functioning under changing environmental conditions (Folke et al. 1996).

In addition to providing human beings with material resources and services, biodiversity has been said to have so-called amenity value (Norton 1987, 99). In other words, biodiversity provides services which are not as such necessary for mere survival, but which can elevate our standard of living and joy for life (Norton 1987, 99; Takacs 1996, 213). A list of this kind of amenity includes, for instance, activities such as bird-watching, hiking, and canoeing. Biodiversity also has a value for science, by providing material for biological study. The variety of life can thus be seen as an intellectual resource, which can satisfy our need to explore our surroundings and can even inspire major scientific discoveries (Takacs 1996, 197–98). It should be noted that amenity value usually depends directly on the presence of relatively unspoiled natural habitats; thus, fulfilling the needs associated with them largely depends on native biodiversity (Norton 1987, 100).

It has been suggested that many of the use and nonuse values of biodiversity can be expressed in monetary terms and thus have economic value. Markets exist for many products based on elements of biodiversity, and for some amenities (e.g., ecotourism). Moreover, several attempts have been made recently to estimate the price of ecosystem services provided by biodiversity. For instance, the average annual economic value of ecosystem services has been estimated as $33 trillion for the entire biosphere (Costanza et al. 1997) and as $319 billion for the United States (Pimentel et al. 1997). Biodiversity or its elements may also have so-called option values, which means that an object that we are unable to exploit now may become valuable in the future (Randall 1988, 84).

Some scientists consider the economic valuation of biodiversity necessary for its conservation, since economic arguments are likely to appeal to decision makers and to the general public (Pearce and Moran 1995, xi–xii; Costanza et al. 1997). However, many biologists and philosophers are skeptical about the idea of placing a monetary value on biodiversity. First, for some elements of biodiversity there is no market, and thus no known market value, making their evaluation in monetary terms impossible. Second, economic valuation methods for expressing current and future values of biodiversity in monetary terms may be untrustworthy and easily biased. We simply lack the knowledge needed for an accurate evaluation of future economic values (Norton 1987, 26). Finally, to assign a monetary value to an object is to treat it as a market commodity. A number of philosophers have argued that biodiversity is among those things which we are not morally entitled to treat as such. According to them, expressing biodiversity value in economic terms is incompatible with the ethically appropriate attitude toward biodiversity (Norton 1987; O'Neill 1997; Gatto and De Leo 2000), which is the subject of the next section.

1.4.2   Intrinsic Value

The term "intrinsic value" has several different meanings (O'Neill 1992, 119). In this chapter, intrinsic value is understood to be synonymous with noninstrumental value. In other words, the intrinsic value of an entity refers to the value that it has in and for its own right, independent of its use, function, or value to any other object (Norton 1987, 7; Callicott 1988, 140; O'Neill 1992, 119; Cox 1997, 174). Accepting the intrinsic value of biodiversity would add a powerful argument for its conservation, since if an entity is perceived as intrinsically valuable, then any moral agent has a moral reason to preserve it (Regan 1988, 200). As a matter of fact, the belief that biodiversity has an intrinsic value appears to be rather widespread among biologists and environmental philosophers (Brennan 1988, 141; Ehrenfeld 1988, 214; Sagoff 1995; Falk, Ehrlich, Noss, Orians, Raven, Soulé, and Wilson according to Takacs 1996; Cafado and Primack 2001, 593; Ghilarov 2000). However, some scientists (e.g., Brussard, Erwin, Janzen, McNaughton in Takacs 1996, 253) deny the intrinsic value of biodiversity. They believe that values are human constructs; if there were no human beings, there would be no values either.

This challenge can be answered in at least two different ways. First, it can be questioned whether value assignments really are human decisions. According to Rolston (1988), humans only discover values but do not create them. The source or generator of the intrinsic value of biodiversity, according to this view, is biodiversity itself (Rolston 1988, 27–8; Lee 1996, 298). The second possibility, theoretically less heavily loaded, is to argue, following Callicott (1988, 142–3), that even though human consciousness is the source of all values, something may be valued for its own sake rather than, or in addition to, for the sake of any subjective experience (pleasure, knowledge, aesthetic satisfaction) it may afford to the valuer. According to this view, all values are human constructs, but this does not imply that there are no intrinsic values. For example, we tend to value newborn infants not only for their instrumental value (for instance the joy they bring to our lives, or the love, help, and care we will receive from them in the future), but also for themselves. In that case, the infant is the locus of values and is intrinsically valuable. In the case of biodiversity, the locus of intrinsic value is thus biodiversity itself. This emphasis on the locus also allows one to resist the inference that only humans can possess intrinsic value (Lee 1996, 298).

The possibility that nonhuman entities may be intrinsically valuable, however, does not necessarily mean that biodiversity is among those entities. The proponents of the intrinsic value of biodiversity therefore need to present further arguments. It has quite commonly been taken for granted that human beings are intrinsically valuable, and the animal rights movement has argued for the intrinsic value of animals by claiming that animals share morally relevant characteristics with human beings. To argue similarly – on the basis of shared characteristics – for the intrinsic value of biodiversity seems very difficult. Some take this to mean that the idea of the intrinsic value of biodiversity remains largely a question of personal belief, and its persuasive power is thus rather low (e.g., Bengtsson et al. 1997). However, many philosophers disagree, asserting that in addition to this argument there are also other, more successful arguments in favor of the intrinsic value of biodiversity. Some theoretically powerful arguments have been presented, for example, by Taylor (1981), Regan (1981), Rolston (1988), Callicott (1988), and Lee (1999) (but see counterarguments by Norton 1984; Cahen 1995). Another, more practical problem with the intrinsic value of biodiversity is that it is sometimes difficult to convince people with theoretical arguments; even theoretically flawless arguments do not always get people to change their views and behavior on biodiversity issues. We discuss the implications of acceptance or rejection of the intrinsic value of biodiversity in section 1.6.2.

1.5   BIODIVERSITY AS A SOCIAL AND POLITICAL CONSTRUCT

From the very beginning, the use of the word biodiversity was not restricted to the academic sphere. Biologists have actively promoted the idea of biodiversity to the public (Takacs 1996), and the term has rapidly been adopted by the media and by politicians. It is important to realize, however, that the meaning of the word biodiversity as used by the media, politicians, and the environmentally aware public differs to some extent from its meaning as a scientific term. In explaining the meaning of the concept of biodiversity to the public, biologists usually take a more intuitive approach and appeal to manifestations or aspects of biodiversity which are most familiar and attractive to people. These aspects, naturally, differ for different people (e.g., charismatic megavertebrates, beautiful landscapes), making the popular version of the concept of biodiversity "fundamentally indefinable" (Swingland 2001, 379). Biodiversity thus becomes "a collective noun that represents a lot of different things" (Eisner, quoted in Takacs 1996, 81). Paradoxically, this imprecision appears to be one of the main reasons for the popularity of the word biodiversity in nonscientific circles:

> If *biodiversity* is blurry and all-encompassing, that is in part why it has been so successful as a conservation buzzword . . . Biodiversity has entered the dictionary, people respond to it, it works, because each of us can find in it what we cherish . . . What is it you most prize in the natural world? Yes, biodiversity is that, too. In biodiversity each of us finds a mirror for our most treasured natural images, our most fervent environmental concerns. (Takacs 1996, 81)

These differences in the meaning of popular and scientific concepts of biodiversity need not be cause for concern, as long as scientists and policy makers are aware of them. The aim of biodiversity as a sociopolitical construct is to draw people's attention to the environmental crisis, to unprecedented extinction rates, and so on; it is "part of a convincing strategy" (Takacs 1996, 79). The term biodiversity is used by biologists "to flag the importance of, and the potential threats associated with, global environmental changes to our evolutionary heritage" (Bowman 1993), and so far it has performed this function well. The differences in the values placed on biodiversity per se and on its elements by scientists and by other members of society may have more serious practical implications. These will be discussed in section 1.6.2.

1.6   PRACTICAL IMPLICATIONS OF BIODIVERSITY DEFINITION AND VALUES

Both the definition of biodiversity and biodiversity values have important implications for biodiversity assessment and conservation strategies (Haila and Kouki 1994). In practice, the development of biodiversity conservation strategies and priorities is the product of interaction between the scientific community and other parts of society (Fig. 1.3). Scientists work with the scientific concept of biodiversity and its operational measures, and argue for various values associated with biodiversity. Both the concept and the values are disseminated to the society (largely through the media) and to decision makers. The resulting popular concept of biodiversity and its values may (as discussed in section 1.5) differ to a greater or lesser extent from the scientific concept. It is the combination of the elements and attributes of biodiversity included in the definition, the values associated with them, and the availability of ways of measuring them which determines the priorities and strategies of biodiversity conservation. Changes in conservation strategies, on the other hand, may require the development of new operational measures. In the following, we examine the role of the biodiversity definition and values in more detail.

1.6.1   Implications of Definition

The scientific definition of biodiversity is a statement of what constitutes the elements of diversity within a biological community. By including or excluding elements from the sphere of biodiversity, we may influence the choices made in biodiversity conservation. The concept of biodiversity also

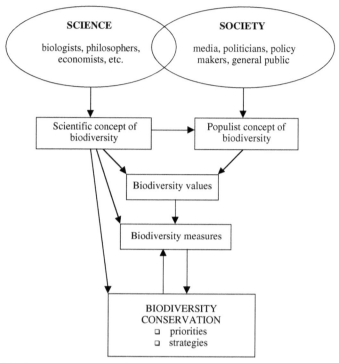

Figure 1.3. Interacting roles of science and society in the development of the concept of biodiversity and in biodiversity conservation

has important implications for conservation practices by revealing the structure and causal relationships among biodiversity components.

The above point can be illustrated by considering the implications of the hierarchical structure of biodiversity. On the one hand, diversities of components at different hierarchical levels are not necessarily correlated even within one (e.g., taxonomic) hierarchy (Angermeier and Karr 1994; Prance 1996). Thus, management designed to maintain one aspect of biodiversity (e.g., species richness) will not necessarily maintain others (e.g., diversity at higher taxonomic levels; Groombridge 1992, xiv). On the other hand, the elements of biodiversity are intimately interlinked and actions at one level may often have an unanticipated impact on other levels. For instance, creating clear-cuts in a forest may seem to increase diversity on a landscape scale, but may reduce diversity at the species level due to the replacement of sensitive species by cosmopolitan weeds (Noss in Takacs 1996, 49). Therefore, either conservation efforts should be directed toward all or most levels of the biological hierarchy, or the conservation program should include a careful definition of the types of biodiversity that are being conserved, and possible side effects on other biodiversity components should be considered.

The assumed causal relationship between the diversity of life forms (entities) and the maintenance of life processes may also affect management and conservation strategies. If the maintenance of ecological processes is considered as the primary means to preserve biodiversity, then species that play a particularly important role for ecosystem functioning – "drivers" (Walker 1992), "keystones" (Paine 1966), or "ecosystem engineers" (Jones et al. 1994) – are likely to become the main focus of conservation efforts. Note that ecological processes do not need to be an integral part of the concept of biodiversity in order to become a primary target of conservation. If, on the other hand, ecosystem functioning is believed to be maintained by the diversity of life forms, the emphasis of conservation should be on maintaining the greatest number of genetically, phylogenetically, and functionally distinct organisms. Of course, the two approaches may not be mutually exclusive, but they clearly prescribe different priorities for conservation work. They would also require the use of different biodiversity measures: indices assessing the functional importance of species would be appropriate in the first case, while measures assessing the degree of phylogenetic and genetic distinctiveness would be more suitable for assessing the diversity of life forms. Once again, the choice of a biodiversity index depends critically on its intended use.

While current conservation practices appear to be at the stage of transition from preservation of the diversity of entities (e.g., endangered species) to preservation of the integrity of biological processes, setting preservation of life processes as the ultimate goal of conservation may be a risky option. For instance, if ecosystem functions could be sustained at a relatively low level of species richness and performed by alien species, making processes the sole target of conservation could lead to a loss of much of the biodiversity native to the area (Soulé 1996; Nott and Pimm 1997; Schwartz et al. 2000). The forests of the Hawaiian Islands offer a graphic example of a normally functioning ecosystem which contains virtually no native tree species (Nott and Pimm 1997). An additional problem with processes as a conservation target is that the a priori identification of the functional importance of different species is methodologically quite difficult (Bond 1993), and there are numerous examples where the keystone status of a species has become evident only after its extinction from the ecosystem. The examination of the relationship between species richness and the functioning of different ecosystems, and the development of methods for identifying functionally important species, are therefore among the most urgent tasks in ecology.

1.6.2   Implications of Values

The answer to the question regarding the kinds of values biodiversity possesses also has very important implications for biodiversity conservation, since value arguments are commonly used by the proponents of conservation. If bio-diversity is found to be intrinsically valuable, we have strong moral reasons to conserve all aspects of biodiversity, regardless of their potential utilitarian and instrumental values. If, conversely, biodiversity is found to be only in-strumentally valuable, then on moral grounds we can demand conservation only of those parts which (directly or indirectly) enhance (or will in the future enhance) the well-being or quality of some other valuable entity or state of affairs.

Since accepting the intrinsic value of biodiversity would provide grounds for preserving all existing biodiversity, many biologists would like to see it as the main argument for biodiversity conservation (e.g., Lawton 1991; Ghilarov 2000). However, as discussed in section 1.4.2, it is difficult to convince people of the arguments for intrinsic value, and thus in practice conservation work rarely rests solely on such value. As a result, conservationists are burdened with the need to find or create instrumental values for each biodiversity ele-ment. This dilemma is clearly revealed in the answers biologists have given when asked why they are personally concerned about biodiversity preserva-tion, and why other members of society should share this concern (Takacs 1996, 277–80). "Human economic and health needs were the most frequently expressed reasons why society should care, but they ranked far down the list of reasons why the biologists themselves cared" (277). Many biologists therefore seem to assume that by emphasizing the economic value of biodi-versity they will have a stronger argument for preserving biodiversity with policy makers and the general public, even though they themselves find argu-ments based on intrinsic and aesthetic values sufficiently convincing (Myers 1996; Bengtsson et al. 1997; cf. also Gamborg and Sandøe in this volume). As Mark Sagoff (1995, 174) puts it: "[w]e environmentalists often appeal to instrumental arguments for instrumental reasons, i.e., not because we believe in them, but because we think that they work." Some biologists feel uneasy about using these double standards (one for the public, another for the sci-entific community), and find it not only dishonest but also dangerous and harmful to biological conservation (Leopold 1970, 247; Lawton 1991; Soulé in Takacs 1996, 283; Ghilarov 1997). The double standard also brings with it the problem of compatibility of standards. As noted in section 1.4.1, the economic value of biodiversity may be incompatible with its intrinsic value because it seems to be morally inappropriate to assign a monetary value to intrinsically valuable objects.

Assigning instrumental value to biodiversity per se by using the 'diversity–stability' and 'biodiversity as insurance' arguments (see section 1.4.1) represents another alternative which would allow the preservation of all (or at least a large number) of the components of biodiversity. Each component of biodiversity could be metaphorically considered as a rivet, playing a small but significant role in keeping the airplane in the air (Ehrlich and Ehrlich 1981). Losing a few rivets will not significantly endanger the flight; but if the process continues, at some point the plane will collapse, and so will the spaceship Earth. The danger is that we do not know at which point the disaster will happen, and it is therefore prudent to preserve "every cog and wheel" (Leopold 1970). This line of reasoning, however, has been criticized as lacking scientific proof, since we do not have sufficient knowledge about the relationship between biodiversity dynamics and ecosystem functioning (Mann 1991; Ghilarov 1996, 1997, 2000; Soulé 1996; Schwartz et al. 2000). The proponents of the precautionary approach in conservation argue, however, that the burden of proof in assessing the consequences of biodiversity loss should be reversed to demonstrate the absence of ecologically significant long-term changes (Dayton 1998), and that arguments for biodiversity conservation based on the link between biodiversity and ecosystem functioning should not be discarded as "weak" but rather should be considered as complementary to other arguments, based on the instrumental or intrinsic value of biodiversity, since they motivate different groups of people to support conservation (Hector et al. 2001).

The question of biodiversity values is also highly relevant to the debate over the place of human-generated diversity in the concept of biodiversity. The reluctance of some scientists to accept human-generated biodiversity as part of the concept may be due in part to the tendency to see in biodiversity values which human-generated elements fail to have. Although many human-generated elements have a high use value, they are usually not valued for the same reasons as natural biodiversity. Moreover, some human-generated elements (for example alien species) are sometimes even disvalued. If we accept these controversial components as a part of biodiversity, we must either give up the intuition that high biodiversity is always valuable or change our mind about the value of the elements. The first alternative means that we must accept that in some cases increases in biodiversity may have no value and biodiversity losses may even be desirable; in other words, we need to accept that only some parts of biodiversity are valuable. The second alternative means that we consider all controversial components as valuable ones and thus as appropriate targets for conservation efforts.

1.7   CONCLUSIONS

Although the term biodiversity has already gained wide popularity in the absence of a unique definition, greater precision is required for purposes of management and conservation (Groombridge 1992; DeLong 1996). There is also the danger that the constant use and misuse of the word biodiversity by the media may cause a negative reaction to the term, leading to the rejection of a serious scientific issue (Walker 1992). In this chapter, we have tried to demonstrate that much of the confusion surrounding the use of the term biodiversity appears to be due to the confounding on the one hand of biodiversity as a scientific concept, along with biodiversity assessment and applications (biodiversity conservation), on the other hand, of biodiversity as a social and political construct (DeLong 1996; Gaston 1996b). This confusion could be minimized by recognizing the differences between these two sides of the phenomenon of biodiversity and specifying the context in which the term is being used (cf. Haila and Kouki 1994).

The concept of biodiversity has also been criticized for its lack of novelty. For instance, Takacs (1996) points out that before the term biodiversity (or biological diversity) was coined, biologists used many other terms and concepts – including nature, wilderness, natural variety, and ecological diversity – which conveyed a very similar message. Ghilarov (1996) has drawn attention to the fact that considerable progress in the study of species and community diversity was already achieved in the 1960s and 1970s. In fact, an interest in the multiplicity of biological kinds can be traced back to Plato, Aristotle, and other philosophers and scientists of classical antiquity (see Juhani Pietarinen's chapter in this volume). In this sense, biodiversity can indeed be seen as "a new linguistic bottle for the wine of old ideas" (Harper and Hawksworth 1996, 5).

The value and novelty of the concept of biodiversity should perhaps not be seen in providing a new entity for study but rather in providing a new approach (Takacs 1996, 75). It forces scientists to think much more broadly than they have done in the past (Noss in Takacs 1996, 49), and it clearly promotes interdisciplinary research. Within biology alone, work on biodiversity links the fields of biogeography, botany, conservation biology, ecology, evolutionary biology, genetics, paleontology, systematics, and zoology (Gaston

1996b). By promoting interdisciplinarity, the concept of biodiversity encourages scientists to attack issues which could not have been addressed within a single discipline. It helps to bridge the gaps between the compositionalist and functionalist approaches (Callicott et al. 1999), and between holistic and reductionist assumptions (Weber and Schmid 1995). A certain progress has been achieved for instance in merging the population- and ecosystem-oriented ecology approaches by studying the links between compositional biodiversity (e.g., species richness) and ecosystem functioning (processes) (Schulze and Mooney 1993; Tartowski et al. 1997; Tilman 1999; Loreau et al. 2001). The strengthening of the links between scientists, environmental philosophers, policy makers, the media, and the public is also very important. Biology in general, and ecology in particular, has been criticized for being a soft science, with poor predictive power (Peters 1991); addressing urgent practical questions on a sound multidisciplinary scientific basis is one way to overcome this crisis (Western 1992).

BIBLIOGRAPHY

Angermeier, P. L. 1994. Does Biodiversity Include Artificial Diversity? *Conservation Biology* 8:600–2.

Angermeier, P. L. and J. R. Karr. 1994. Biological Integrity versus Biological Diversity as Policy Directives: Protecting Biotic Resources. *BioScience* 44:690–97.

Bengtsson, J., H. Jones, and H. Setälä. 1997. The Value of Biodiversity. *Trends in Ecology and Evolution* 12:334–6.

Bond, W. J. 1993. Keystone Species. In *Biodiversity and Ecosystem Function*, E.-D. Schultze and H. A. Mooney, eds. Berlin: Springer-Verlag.

Bowman D.M.J.S. 1993. Biodiversity: Much More Than Biological Inventory. *Biodiversity Letters* 1:163.

Brennan, A. 1988. *Thinking about Nature: An Investigation of Nature, Value and Ecology*. London: Routledge.

Cafado, P. J. and R. B. Primack. 2001. Ethical Issues in Biodiversity Protection. In *Encyclopedia of Biodiversity*. Vol. 2, S. A. Levin, ed. San Diego: Academic Press.

Cahen, H. 1995. Against the Moral Considerability of Ecosystems. In *People, Penguins, and Plastic Trees: Basic Issues in Environmental Ethics*, C. Peirce and D. VanDeVeer, eds. Belmont: Wadsworth Publishing.

Callicott, J. B. 1988. On Intrinsic Value of Nonhuman Species. In *The Preservation of Species: The Value of Biological Diversity*, B. G. Norton, ed. Princeton, N.J.: Princeton University Press.

Callicott, J. B., L. B. Crowder, and K. Mumford. 1999. Current Normative Concepts in Conservation. *Conservation Biology* 13:22–35.

Costanza, R., R. d'Arge, R. de Groot, S. Farber, M. Grasso, B. Hannon, K. Limburg, S. Naeem, R. V. O'Neill, J. Paruelo, R. G. Raskin, P. Sutton, and M. van den Belt. 1997. The Value of the World's Ecosystem Services and Natural Capital. *Nature* 387:253–60.

Cousins, S. H. 1991. Species Diversity Measurement: Choosing the Right Index. *Trends in Ecology and Evolution* 6:190–2.

Cox, D. 1997. On the Value of Natural Relations. *Environmental Ethics* 19:173–83.

Dasmann, R. F. 1991. The Importance of Cultural and Biological Diversity. In *Biodiversity: Culture, Conservation, and Ecodevelopment*, M. L. Oldfield and J. B. Alcorn, eds. Boulder, Colo.: Westview Press.

Dayton, P. K. 1998. Reversal of the Burden of Proof in Fisheries Management. *Science* 279:821–2.

DeLong, D. C. 1996. Defining Biodiversity. *Wildlife Society Bulletin* 24:738–49.

Ehrenfeld, D. 1988. Why Put Value on Biodiversity. In *BioDiversity*, E. O. Wilson, ed. Washington, D.C.: National Academy Press.

Ehrlich, P. and A. Ehrlich. 1981. *Extinction: the Causes and Consequences of the Disappearance of Species*. New York: Random House.

Erwin, T. L. 1991. An Evolutionary Basis for Conservation Strategies. *Science* 253: 750–2.

Folke, C., C. S. Holling, and C. Perrings. 1996. Biological Diversity, Ecosystems, and the Human Scale. *Ecological Applications* 6:1018–24.

Franklin, J. F. 1988. Structural and Functional Diversity in Temperate Forests. In *BioDiversity*, E. O. Wilson, ed. Washington, D.C.: National Academy Press.

Gaston, K. J., ed. 1996a. *Biodiversity: A Biology of Numbers and Difference*. Oxford: Blackwell Science.

1996b. What Is Biodiversity? In *Biodiversity: A Biology of Numbers and Difference*, K. J. Gaston, ed. Oxford: Blackwell Science.

1996c. Species Richness: Measure and Measurement. In *Biodiversity: A Biology of Numbers and Difference*, K. J. Gaston, ed. Oxford: Blackwell Science.

Gaston, K. J. and J. I. Spicer. 1998. *Biodiversity: An Introduction*. Oxford: Blackwell Science.

Gatto, M. and G. A. de Leo. 2000. Pricing Biodiversity and Ecosystem Services: The Never-ending Story. *BioScience* 50:347–55.

Ghilarov, A. M. 1996. What Does 'Biodiversity' Mean – Scientific Problem or Convenient Myth? *Trends in Ecology and Evolution* 11:304–6.

1997. Species Redundancy versus Non-Redundancy: Is It Worth Further Discussion? *Zhurnal Obshchej Biologii* 58:100–105.

2000. Ecosystem Functioning and Intrinsic Value of Biodiversity. *Oikos* 90:408–12.

Glowka, L., F. Burhenne-Guilmin, H. Synge, J. A. McNeely, and L. Gündling. 1994. *A Guide to the Convention on Biological Diversity*. Gland and Cambridge: IUCN.

Groombridge, B., ed. 1992. *Global Biodiversity: Status of the Earth's Living Resources*. A report compiled by the Word Conservation Monitoring Centre. London: Chapman and Hall.

Haila, Y. and J. Kouki. 1994. The Phenomenon of Biodiversity in Conservation Biology. *Annales Zoologici Fennici* 31:5–18.

Harper, J. L. and D. L. Hawksworth. 1996. Preface. In *Biodiversity: Measurement and Estimation*, D. L. Hawksworth, ed. London: Chapman and Hall.

Hector, A., J. Joshi, S. P. Lawler, E. M. Spehn, and A. Wilby. 2001. Conservation Implications of the Link Between Biodiversity and Ecosystem Functioning. *Oecologia* 129:624–8.

Heywood, V. H. and I. Baste. 1995. Introduction. In *Global Biodiversity Assessment*, V. H. Heywood and R. T. Watson, eds. Cambridge: Cambridge University Press.

Hughes, J. B., G. C. Daily, and P. R. Ehrlich. 1997. Population Diversity: Its Extent and Extinction. *Science* 278:689–92.

Hurlbert, S. H. 1978. The Measurement of Niche Overlap and Some Relatives. *Ecology* 59:67–77.

ICBP 1992. *Putting Biodiversity on the Map: Priority Areas for Global Conservation.* Cambridge: ICBP (BirdLife International).

Jeffries, M. J. 1997. *Biodiversity and Conservation.* London: Routledge.

Jones, C. G., J. H. Lawton, and M. Shachak. 1994. Organisms as Ecosystem Engineers. *Oikos* 69:373–86.

Lawton, J. H. 1991. Are Species Useful? *Oikos* 62:3–4.

1994. What Do Species Do in Ecosystems? *Oikos* 71:367–74.

Lee, K. 1996. The Source and Locus of Intrinsic Value: A Reexamination. *Environmental Ethics* 18:297–309.

1999. *The Natural and the Artefactual: The Implications of Deep Science and Deep Technology for Environmental Philosophy.* Lanham: Lexington Books.

Leopold, A. 1970. *A Sand County Almanac.* San Francisco: Sierra Club.

Loreau, M., S. Naeem, P. Inchausti, J. Bengtsson, J. P. Grime, A. Hector, D. U. Hooper, M. A. Huston, D. Raffaelli, B. Schmid, D. Tilman, and D. A. Wardle. 2001. Biodiversity and Ecosystem Functioning: Current Knowledge and Future Challenges. *Science* 294:804–8.

Lovejoy, T. E. 1980. Changes in Biological Diversity. In *The Global 2000 Report to The President, vol. 2 (The Technical Report)*, G. O. Barney, ed. Harmondsworth: Penguin Books.

McCann, K. S. 2000. The Diversity-Stability Debate. *Nature* 405:228–33.

Magurran, A. E. 1988. *Ecological Diversity and Its Measurement.* London: Croom Helm.

Mallet, J. 1996. The Genetics of Biological Diversity: From Varieties to Species. In *Biodiversity: A Biology of Numbers and Difference*, K. J. Gaston, ed. Oxford: Blackwell Science.

Mann, C. C. 1991. Extinction: Are Ecologists Crying Wolf? *Science* 253:736–38.

Myers, N. 1996. Environmental Services of Biodiversity. *Proceedings of the National Academy of Sciences of USA* 93:2764–9.

1997. The Rich Diversity of Biodiversity Issues. In *Biodiversity II*, M. L. Reaka-Kudla, D. E. Wilson, and E. O. Wilson, eds. Washington, D.C.: Joseph Henry Press.

Norton, B. G. 1984. Environmental Ethics and Weak Anthropocentrism. *Environmental Ethics* 6:131–48.

1987. *Why Preserve Natural Variety?* Princeton, N.J.: Princeton University Press.

Noss, R. F. 1990. Indicators for Monitoring Biodiversity: A Hierarchical Approach. *Conservation Biology* 4:355–64.

1996. Ecosystems as Conservation Targets. *Trends in Ecology and Evolution* 11:351.

Nott, M. P., and S. L. Pimm. 1997. The Evaluation of Biodiversity as a Target for Conservation. In *The Ecological Basis for Conservation: Heterogeneity, Ecosystems and Biodiversity*, S. T. A. Pickett, R. S. Ostfeld, M. Shachak, and G. E. Likens, eds. New York: Chapman and Hall.

Office of Technology Assessment (OTA). 1987. *Technologies to Maintain Biological Diversity*. Washington, D.C.: U.S. Government Printing Office.

O'Neill, J. 1992. The Varieties of Intrinsic Value. *The Monist* 75:119–33.

———. 1997. Managing Without Prices: The Monetary Valuation of Biodiversity. *Ambio* 26:546–50.

Paine, R. T. 1966. Food Web Complexity and Species Diversity. *American Naturalist* 100:65–75.

———. 1992. Food Web Analysis through Field Measurement of Per Capita Interaction Strength. *Nature* 355:73–5.

Pearce, D. and D. Moran. 1995. *The Economic Value of Biodiversity*. London: Earthscan.

Perlman, D. L. and G. Adelson. 1997. *Biodiversity: Exploring Values and Priorities in Conservation*. Cambridge: Blackwell Science.

Peters, R. H. 1991. *A Critique for Ecology*. Cambridge: Cambridge University Press.

Pielou, E. C. 1975. *Ecological Diversity*. New York: Wiley.

Pimentel, D., C. Wilson, C. McCullum, R. Huang, P. Dwen, J. Flack, Q. Tran, T. Saltman, and B. Cliff. 1997. Economic and Environmental Benefits of Biodiversity. *BioScience* 47:747–57.

Prance, G. T. 1996. A Comparison of the Efficacy of Higher Taxa and Species Number in the Assessment of Biodiversity in the Neotropics. In *Biodiversity: Measurement and Estimation*, D. L. Hawksworth, ed. London: Chapman and Hall.

Randall, A. 1988. Human Preferences, Economics, and the Preservation of Species. In *The Preservation of Species: The Value of Biological Diversity*, B. G. Norton, ed. Princeton, N.J.: Princeton University Press.

Regan, T. 1981. The Nature and Possibility of an Environmental Ethic. *Environmental Ethics* 3:19–34.

Regan, D. H. 1988. Duties of Preservation. In *The Preservation of Species: The Value of Biological Diversity*, B. G. Norton, ed. Princeton, N.J.: Princeton University Press.

Rolston, H. 1988. *Environmental Ethics: Duties to and Values in The Natural World*. Philadelphia: Temple University Press.

Sagoff, M. 1995. Zuckerman's Dilemma: A Plea for Environmental Ethics. In *People, Penguins, and Plastic Trees: Basic Issues in Environmental Ethics*, C. Peirce and D. VanDeVeer, eds. Belmont: Wadsworth Publishing.

Samson, F. B. and F. L. Knopf. 1994. A Framework to Conserve Biological Diversity through Sustainable Land Management. *Transitions of the Northern American Wildlife and Natural Resources Conference* 59:367–77.

Savage, J. M. 1995. Systematics and the Biodiversity Crisis. *BioScience* 45:673–9.

Schläpfer, F. and B. Schmid. 1999. Ecosystem Effects of Biodiversity: Classification of Hypotheses and Exploration of Empirical Results. *Ecological Applications* 9:893–912.

Schulze, E. D. and H. A. Mooney, eds. 1993. *Biodiversity and Ecosystem Function*. New York: Springer-Verlag.

Schwartz, M. W., C. A. Brigham, J. D. Hoeksema, K. G. Lyons, M. H. Mills, and P. J. van Mantgem. 2000. Linking Biodiversity to Ecosystem Function: Implications for Conservation Ecology. *Oecologia* 122:297–305.

Schwarz, C. F., E. C. Thor, and G. H. Elsner. 1976. *Wildland Planning Glossary*. USDA Forest Service General Technical Report PSW-13, Pacific Southwest Forest and Range Experimental Station, Berkeley, California.

Sober, E. 1988. Philosophical Problems for Environmentalism. In *The Preservation of Species: The Value of Biological Diversity*, B. G. Norton, ed. Princeton, N.J.: Princeton University Press.

Soulé, M. E. 1996. Are Ecosystem Processes Enough? *Wild Earth* 6:59–60.

Spellerberg, I. F. 1997. Themes, Terms and Concept. In *Conservation Biology*, I. F. Spellerberg, ed. Singapore: Longman.

States J. B., P. T. Haug, and T. G. Shoemaker. 1978. *A System Approach to Ecological Baseline Studies.* FWS/OBS-78/21. Washington, D.C.: U.S. Fish and Wildlife Service.

Swingland, I. R. 2001. Biodiversity, Definition of. In *Encyclopedia of Biodiversity*, Vol. 1, S. A. Levin, ed. San Diego: Academic Press.

Takacs, D. 1996. *The Idea of Biodiversity: Philosophies of Paradise.* Baltimore and London: Johns Hopkins University Press.

Tartowski, S. L., E. B. Allen, N. E. Barrett, A. R. Berkowitz, R. K. Colweel, P. M. Groffman, J. Harte, H. P. Possingham, C. M. Pringle, D. L. Strayer, and C. R. Tracy. 1997. Integration of Species and Ecosystem Approaches to Conservation. In *The Ecological Basis for Conservation: Heterogeneity, Ecosystems and Biodiversity*, S. T. A. Pickett, R. S. Ostfeld, M. Shachak, and G. E. Likens, eds. New York: Chapman and Hall.

Taylor, P. W. 1981. The Ethics of Respect for Nature. *Environmental Ethics* 3:197–218.

Tilman, D. 1999. The Ecological Consequences of Changes in Biodiversity: A Search for General Principles. *Ecology* 80:1455–74.

2000. Causes, Consequences and Ethics of Biodiversity. *Nature* 405:208–11.

Vane-Wright, R. I., C. J. Humphries, and P. H. Williams. 1991. What to Protect: Systematics and the Agony of Choice. *Biological Conservation* 55:235–54.

Walker, B. H. 1992. Biodiversity and Ecological Redundancy. *Conservation Biology* 6:18–23.

Weber, M. and B. Schmid. 1995. Reductionism, Holism, and Integrated Approaches in Biodiversity Research. *Interdisciplinary Science Reviews* 20:49–60.

Western, D. 1992. The Biodiversity Crisis: A Challenge for Biology. *Oikos* 63:29–38.

Wheeler, Q. D. 1995. Systematics and Biodiversity: Politics at Higher Levels. *BioScience* 45:21–8.

Whittaker, R. H. 1977. Evolution of Species Diversity in Land Communities. In *Evolutionary Biology*, Vol. 10, M. K. Hecht, W. C. Steere, and B. Wallace, eds. New York: Plenum Press.

Williams, P. H. and C. J. Humphries. 1996. Comparing Character Biodiversity among Biotas. In *Biodiversity: A Biology of Numbers and Difference*, K. J. Gaston, ed. Oxford: Blackwell Science.

Williams, P. H., C. J. Humphries, and K. J. Gaston. 1994. Centres of Seed-plant Diversity: The Family Way. *Proceedings of the Royal Society of London B* 256:67–70.

Wilson, E. O., ed. 1988. *BioDiversity.* Washington, D.C.: National Academy Press.

2

Making the Biodiversity Crisis Tractable

A Process Perspective

YRJÖ HAILA

2.1 INTRODUCTION

In current conservation parlance, "biodiversity" has become a generic term for everything that is good and worth preserving in living nature. It was originally used as a rallying cry of conservation biologists/ecologists/conservationists in the 1980s; Takacs (1996) describes the process, starting from the arrangement of the Forum on BioDiversity in Washington, D.C. in 1986. After the UN Conference on Environment and Development in Rio de Janeiro in 1992, biodiversity became a staple theme in international politics. Parallel to this institutional solidification of biodiversity on the political scene, the term itself has acquired more and more scientific credibility. E. O. Wilson's widely acclaimed monograph (Wilson 1992) was important for making the literary public familiar with the term. The most ambitious scientific project to date is headed by the Princeton ecologist Simon A. Levin, aiming at an *Encyclopedia of Biodiversity*; a five-volume printed version was published in 2001 (Academic Press), and a much larger electronic version is under preparation.

The preceding paragraph summarizes briefly the history of biodiversity as an environmental issue. The term broke through into scientific, political, and public consciousness remarkably quickly. It was invented in the early 1980s by a group of ecologists and evolutionary biologists, basically as a political slogan (see Takacs 1996). In other words, biodiversity rose to the position of an important environmental issue through deliberate social construction (Hannigan 1995; Haila 1999a).

The term "social construction" should not be taken as a simple pejorative; rather, it brings up important aspects of the constitution of the human social world. In the human world, issues and concerns are expressed through concepts, and concepts are articulated in language. The construction of a new concept requires supportive *framing*: the concept has to be embedded in a network of other concepts, which together define its meaning. As Hacking (1999) notes, the attribute "social" is redundant: processes taking place in society are social, by definition. Hacking would reserve the term "construct" for such concepts that are deliberately brought forth, mainly for critical purposes (5–7). This, as we saw, is true of the term biodiversity.

Public space is a useful term for clarifying the process of problem construction. Commonly shared problems and concerns exist in a public space, created by communication. The construction of an issue means that it gains weight and fills an opening in public space. The emergence of problems in public space is a dynamic process. A commonly shared concern may become an organizing center, which influences the perception of other issues as well. This was an explicit aim of the biologists who organized the 1986 Washington meeting (Takacs 1996).

How social problems get constructed has been amply discussed, so I will not dwell on this here; for an analysis of biodiversity, see Hannigan (1995). Instead, I am interested in how the process of construction has fed back to the understanding of the issue itself. Such a feedback loop is relevant as the construction of problems is a continuous process: constant work is needed to keep a particular problem in public consciousness. One can presume that particular issue articulations become stabilized due to contingent events, that careers of issues are "path-dependent." I address this possibility in the last section.

Social issues become material actors in society by mobilizing people. This happens directly when people get motivated to start actions – protest movements, demonstrations, civil disobedience, and the like – and indirectly when a particular issue molds the social space in which human activity takes place. Thus, figuratively speaking, every particular issue has a characteristic *scale*. The scale depends on the "volume" of the public space, which it molds and within which it resides, exerting dominant resonance with other issues (Haila 2002b).

Biodiversity has been framed as a very "big" issue in the most literal sense. It is perceived as being connected to everything – indeed, as a synonym of everything. As Wilson (1997, 1) writes on what biodiversity is: "So, what is it? Biologists are inclined to agree that it is, in one sense, everything." Thus, biodiversity is as good an example as one can wish of an issue with strong resonance effects with other issues. However, making an issue too "big" may also be counterproductive: if an issue covers "everything," how can it simultaneously acquire analytic clarity and strength? In this chapter I explore the tension between the big normative mission and empirical understanding.

Takacs (1996)[1] has laid the groundwork on how the issue was originally launched. I follow the later trajectory of the concept as a dynamic process. As biodiversity research has become a major growth industry, a complete coverage of the relevant literature would be a pipe dream. Instead, I have used as sources major overviews such as topical essay collections and articles in *Nature* and *Science*. The most important sources are, of course, the two authorized anthologies, *Biodiversity* and *Biodiversity II* (Wilson and Peter 1988; Reaka-Kudla et al. 1997, respectively), as well as Wilson's (1992) popular monograph.

2.2   THE DYNAMICS OF FRAMING

Framing an issue means defining a stable context within which it can be adequately understood and addressed. The context allows specified questions to be asked as well as criteria to be given on what could count as an answer to those questions. In scientific work, the stabilization of a research context depends on two prerequisites (see Dyke 1988; Haila 1998).

First, a suitable, or "doable" research procedure must become established. To specify this point, we can use Hacking's (1992) idea that successful research traditions "produce a sort of self-vindicating structure that keeps them stable" (29–30). Such a self-vindicating structure comprises three elements, namely, (1) ideas, (2) things, and (3) marks and manipulation of marks. It is normally only as a result of a laborious process that these elements begin to fit together and support each other. Andrew Pickering's (1995) fitting metaphor for this process of mutual adjustment is *mangle*.

Second, a relevant *problem space* has to be defined. Alan Garfinkel (1981) originally suggested that a problem space is analogous to a physical state space: the "movement" of problem definitions and perceptions within that space is influenced by critical variables, which bound the space. Successful explanation of a problem within a certain problem space requires, in its turn, *explanatory closure* (Dyke 1988, 41–2). In the most elementary case, explanatory closure is achieved when only two alternative explanations are available: refuting one immediately lends support to the other. This corresponds to the paradigmatic ideal of hypothesis testing: construct situations such that observed results unambiguously either refute or support predictions. However, an either-or closure is rare and strictly relative to the problem space upon which it is built. Problem space and closure clearly belong together in principle. However, as Ian Hacking in particular has shown, it may be no trivial task to fit them together in practice. For applications of the state space analogy, see Dyke (1993) and Haila (2002a).

The framing of social problems takes place in two dimensions, as it were: on the one hand, the *normative urgency* of the problem has to be grounded, and on the other hand, the *analytic features* of the problem have to be specified. One could call the first dimension political framing and the second, conceptual framing. The dimensions, of course, interact with each other. The latter dimension also implies an idea on how the problem might be solved, and this cannot be normatively neutral (Haila and Levins 1992, 226). In these terms, the Washington BioDiversity Forum of 1986 was a stage set deliberately for political framing.

As we already noted, framing is a dynamic process. The process aims at defining an adequate problem space, which would allow explanatory closure for empirical research. In the case of biodiversity research, this has proven complicated; I focus here on these difficulties. I have divided the exploration into four thematic sections: *urgency, entities, processes,* and *human subsistence.* These sections correspond to four different framings, which have been used to define and address the biodiversity issue. Each one of them, of course, also builds upon earlier concerns that were originally totally unconnected to the current biodiversity issue. This is more than natural, considering that biodiversity is perceived as synonymous for "everything" in living nature: people have been interested in biology since long before biodiversity was launched. As the temporal span of the whole process is less than two decades, the framings have existed side by side, and often several of them are mingled together in single analyses. I have come to think that the sections present a logical sequence, as will become apparent later.

Two particular aspects have interfered with the framing of biodiversity all along. First, the term biodiversity carries strong normative connotations. Of course, this is no problem in itself, as every important concept has them. However, normative connotations lie on hidden premises, which may carry hidden constraints into the shaping of the problem space through the back door, so to speak. The problem space determines what kind of questions can be legitimately asked. My first aim is to analyze the normative premises that have influenced the analytic understanding of biodiversity. A specific problem is raised by the "bigness" of the biodiversity concern in normative terms: whenever somebody adds qualifications to the normative statements, this may be experienced as an attack against the concern itself.

Second, stabilization of empirical research on biodiversity has proven difficult. Again, considering the broad scope of the issue, this is no wonder: how do you stabilize research on "everything"? My second focus is on the interference of the normative baggage with specified research. Such interference has surfaced as tensions within and gaps between the four framings. In other words, the four framings are not quite concordant with each other.

What I name gaps are really underdeveloped problem areas in which more conceptual and/or empirical work is needed. By naming them gaps I do not suggest that no adequate work has been done. Excellent specific research is being done on many facets of the biodiversity issue. I am not aiming at an overarching criticism of biodiversity research, which would be an impossible task anyway. My goal is much more focused: I explore the tension between normative assertions and empirical research, using the four framings and their mutual relationships as my conceptual structure.

2.3 THE URGENCY

2.3.1 Extinction Crisis

A characteristic feature of biodiversity as an environmental issue is its huge size: nothing less is on the agenda than the future of human civilization. Such "bigness" of perceived problems is nothing new in environmentalist thinking; just consider the term "eco-catastrophe" coined by Paul Ehrlich in his famous popular article with the same title, published in *Ramparts* in 1969. However, biodiversity gives a novel flavor to this perception by giving it a scientific footing. Whereas "eco-catastrophe" was explicitly a social dystopia, built upon hypothetical competition between the super powers to develop ever more efficient insecticides, the biodiversity crisis is about the future of the basic ecological mechanisms maintaining human life on Earth. The mechanisms are considered threatened by normal human subsistence activities.

The interviews in Takacs (1996) reflect the sense of crisis that set into motion the organization of the 1986 conference in Washington, D.C. One of the more extreme formulations is in the published version of Paul Ehrlich's talk at the conference:

> What then will happen if the current decimation of organic diversity continues? . . . As ecosystems services falter, mortality from respiratory and epidemic disease, natural disasters, and especially famine will lower life expectancies to the point where cancer (largely a disease of the elderly) will be unimportant. Humanity will bring upon itself consequences depressingly similar to those expected from a nuclear winter. Barring a nuclear conflict, it appears that civilization will disappear some time before the end of the next century – not with a bang but with a whimper. (Ehrlich 1988, 25)

Quite frankly, Ehrlich's forecast is incomprehensible, but let's leave this aside for the moment. The phrase grounding the forecast, "current decimation of organic diversity," refers to species extinctions. As Takacs (1996) narrates in detail, biologists had become seriously worried about an increasing extinction threat in the 1970s. Early summaries include the books by Myers (1979) and Ehrlich and Ehrlich (1981). Extinctions were the explicit focus of the BioDiversity forum in Washington. Strong words are commonly used to characterize the threat; in his popular monograph Wilson (1992) used expressions such as "miniature holocausts" (p. 265) and "a problem of epic dimensions" (p. 297).

At this point, we have to clarify the argument on extinction threat. There is no doubt that global extinction rates are at present higher than the natural long-term average because of human-induced environmental changes. The first clarification to make, however, is that the notion of natural background is, in itself, a highly abstract notion, as the frequency of extinctions has varied greatly because of major extinction waves in the past (Raup 1988; Jablonski 1995). A more adequate baseline for the comparison would be prehuman extinction rate. But this is problematic too. What time span should be covered? What parts of the world should be pooled together? The latter question is intriguing from the perspective of those parts of the Northern Hemisphere that were covered by continental ice as recently as 10,000 years ago, for instance northwestern Europe. At that time, these regions were totally devoid of higher life-forms. "Prehuman extinction rate" hardly makes sense under such circumstances.

More importantly, the actual estimate of the current extinction rate is controversial, too. Originally, the concern over a human-caused extinction wave was aroused by theoretical predictions derived from the theory of island biogeography (MacArthur and Wilson 1967). According to the theory, the number of species in a particular region such as an island is determined as an equilibrium between colonization and extinction (or speciation and extinction, in evolutionary time). Hence, if the human modification of natural habitats leads to a diminishing total area, an increasing extinction rate follows. The application of the theory to conservation rests on an analogy between islands and preserves, assumed to be surrounded by habitat types that are inhospitable to the majority of the original species. Wilson (1992, 208–16) summarizes the argument. Preston (1962) originally suggested the analogy between preserves and islands; for a history of the argument, see Simberloff (1997).

The use of species–area predictions, derived from island biogeography, to estimate extinction risks brings forth three big problems, which I only mention in this context (see Haila 1990, 1999b, 2002a):

First, the species–area estimates assume that the habitat outside of the preserves is nothing but ecological desert, but this is never quite true. The question to what extent the analogy between habitat fragments and ecologically isolated islands is adequate is an empirical one.

Second, species are not evenly distributed across regions, landscapes, and habitats. Quite the contrary, biodiversity is concentrated into "hotspots" on the geographical scale, and the number of species varies greatly across habitats on regional and local scales. Consequently, quite similar habitats in the same biogeographic zone differ as to their conservation value (for the boreal zone, see Järvinen and Väisänen 1979; Haila and Järvinen 1990; Haila 1994). Several authors have emphasized that this holds true of tropical forests as well. In the proceedings of the Washington Forum, Lugo (1988) made this point very emphatically and with good empirical backing, but his conclusions have not been assimilated into the discussion at all.[2]

Third, the recurring reference to global extinction statistics gives the impression that the driving forces of extinctions are also global. This is an example of what might be called *a fallacy of global averages*. The crux of the matter is as follows: the global biota does not form one functionally unified whole, in which all species would be in an equal position in the way global extinction statistics suggest. A great majority of the known extinctions have, for instance, happened on oceanic islands. For birds, Steadman (1997) gives the following estimates: of 108 birds known to have become extinct since A.D. 1600, 97 (90%) lived on islands.[3] It is a tragedy that these species are gone, but the 1 percent reduction in the total number of bird species on Earth does not mean that every community of birds on every continent would have lost one-hundredth of its vigor. The ethical concern and ecological concern simply do not translate directly into one another.

The perception of an imminent extinction crisis became a major motivating force for biologists to get engaged in conservation issues in the 1980s. As a consequence, conservation biology was established as a new scientific discipline. One of the originators of the discipline, Michael Soulé, described it as a *mission- or crisis-oriented discipline* by writing: "(i)ts relation to biology, particularly ecology, is analogous to that of surgery to physiology and war to political science" (Soulé 1985). The emphasis on extinctions also found resonance among the general public. Conservation organizations had already used rare emblematic animals such as the Giant Panda as their symbols for some time. Furthermore, whale and seal hunting became big issues in the media in the 1970s; Einarsson (1993) analyzes the cultural metaphors used in the anti-whaling campaign of the 1970s. These developments meant that there already was fertile soil for extinction concerns in the public sphere.

2.3.2   Human Population Growth

To make the feeling of crisis real, its driving force had to be specified, too. So, what are the causes of the extinction threat? One standard answer is given to this question, in a variety of formulations: human population growth and increasing human encroachment on the natural world. In the Washington Forum, Paul Ehrlich (1988, 21) gave this explanation: "The primary cause of the decay of organic diversity is . . . the habitat destruction that inevitably results from the expansion of human population and human activities." In the second conference, Thomas Lovejoy (1997, 12) made the point as follows: "Beyond the immediate causes that threaten biodiversity, there are ultimate causes such as human population growth . . . and the massive impact of economic activities." Wilson's (1992, 260) words are as follows: "Human demographic success has brought the world to this crisis of biodiversity."

We need to take another brief pause to consider this argument. To begin with, let's make clear what it means to identify the *cause of a crisis*. This is a dual concept. On the one hand, the cause is clearly an explanation for the crisis in question: it points out the factors that brought the crisis about. On the other hand, the cause also implies what should be done: it brings into focus such factors that ought to be modified or removed in order that the crisis be avoided. The argument that human population growth is the cause of the biodiversity crisis is unpersuasive on both counts.

First, the explanatory aspect: if we accept human population growth as *the cause* of a looming extinction crisis, we lose the ability to draw distinctions between more threatening and less threatening situations. We meet the fallacy of global averages once again: human population growth is considered on the level of global average, and extinction threat becomes a global average as well. To demonstrate the logic, I take a somewhat lengthy citation from Ehrlich:

1. Rates of extinctions of both populations and species are related to the rate of habitat loss.
2. The rate of habitat loss increases with the scale of the human enterprise.
3. Total energy use is a reasonable, if imperfect, surrogate for the scale of the human enterprise and its environmental impact. . . . [A] doubling of energy use leads to roughly a doubling of the rate. (1995, 215)

The logic in this citation is simply flawed. Human populations have grown in all inhabited parts of Earth during the last centuries, but the biodiversity crisis is not everywhere the same. Ehrlich's argument builds upon an implicit assumption that Earth is a unified system as far as the human population goes and that, consequently, the ecological effects of human activity can be pooled together into one single figure on the global level. In other words, the assumption is that the "carrying capacity" of the global environment for humans can be assessed within a Malthusian closure. This is patently incorrect: both technological and social innovations have brought forth "free lunches" throughout human history, and human influence on the environment is not a monotonously increasing function of population size or economic productivity, as many economic historians have pointed out (e.g., Mokyr 1990). In particular, increasing use of energy has facilitated enormous increases in the efficiency of resource use and, in many cases, decrease in the rate of habitat loss. The history of forest use in northwestern Europe and the eastern United States are good examples. Slash-and-burn agriculture was the prevailing mode of forest use in vast areas in northern Europe until the late nineteenth century, and it subjected the forest ecosystem to much heavier stress than modern forest management (Haila and Levins 1992). The waves of deforestation and then reforestation of the eastern United States are described in Michael Williams (1989).

Malthusian closure implies that the relationship between human subsistence systems and the productivity of other ecological systems is a zero-sum game, but this is not adequate either. On the global level, it is simply impossible to do such thermodynamic calculations that would allow an assessment of this point (Morowitz 1992; Haila 1999c). Regionally and locally there is huge variation in natural conditions across Earth in this regard. Northwestern Europe, which got its extant flora and fauna after the latest glaciation practically simultaneously with human inhabitants is in a completely different situation than the ancient Gondwanan flora and fauna of West Australia (Hopper et al. 1996).[4]

Second, the practical relevance: if population growth, or growth of the human enterprise, is the only explanation we can give to the loss of biodiversity, practical suggestions on what should be done wear thin. Let's not misunderstand this point. It definitely is an important goal to get the human population to level off, the sooner the better. But the point is again that we need to make distinctions. We should focus on how things are done, and on particular ways of using nature in specific situations. This is often neglected in scenarios of global resource use even in situations where consideration of alternative ways of using – or not misusing – particular resources almost compel themselves.[5] A mere focus on population size, or the size of the economy, loses sight of this point. This is all the more important because economy is not a "thing," it is more adequately envisaged as a complex of cyclic flows. Rates and rhythms matter more than the total volume (see Dyke 1981; Jacobs

2000). Consequently, there is no simple way, or perhaps no way at all, of constructing an accounting system, which would be based on entropy or any other relevant physical characteristic and cover the whole economy (Dyke 1992, 1994). Calculations based on market prices are totally inadequate for these purposes.

The brutal, practical bottom line is that if we only focus on the human population, we can hardly come up with other remedies than to curb population growth. This raises uncomfortable questions, such as: If surgery is an appropriate metaphor for the "crisis discipline" of conservation biology, as Soulé (1985) suggests, what is the tumor that has to be removed?[6]

To summarize, the crisis framing of biodiversity has support from strong underlying normative sentiments. They emphasize the urgency of the crisis, using powerful metaphors such as nuclear winter and the holocaust, and point toward the growth of the human population as the ultimate cause of the crisis. My conclusion is that neither of these claims suggests guidelines for specified hypotheses or conclusions. In other words, the crisis framing has not created fruitful interaction with the needs of practical research.

Furthermore, naming global human population growth as the cause of the crisis is counterproductive because it merges together all types of human influence on the natural world. It supports a total dualistic view about humanity–nature relationships (this argument is elaborated in Haila 1999a, 2000). It also leaves completely unanswered the question of sustainable subsistence practices; I return to this point later.

2.4   ENTITIES

2.4.1   Instrumental Needs

To do anything about biodiversity, we need to know where it is and in what particular shapes it exists. Instead of trying to address a general overwhelming concern, we need practical ways to identify and measure biodiversity. At the very least, we should be able to construct scales that allow comparisons and prioritization in specific contexts. For this purpose, we need clearly identifiable indicators or surrogates of biodiversity,[7] and methodological skills to survey and monitor them using quantitative approaches. Essays in Usher (1986) and Margules and Austin (1991) articulate clearly the practical needs; Paul Williams (1998) gives a good overview of the development of this research field. The task is all the more daunting because of the lack of adequate databases even to monitor extinctions, let alone to track less conspicuous ecological changes (Margules and Austin 1995).

The need to name specific surrogates for biodiversity was, of course, acknowledged early on. The semi-official definition of biodiversity in political documents such as the Rio convention emphasizes entities, too, and thus lends support to the practical needs. In political documents, biodiversity is usually divided into three main levels of biological organization, namely, the diversity of genes in populations, the diversity of populations in ecosystems, and the diversity of ecosystems in landscapes and biogeographic regions. In other words, biodiversity is manifest in the heterogeneity (variability) of biological entities making up systems on different levels of biological hierarchy.

Assessment of global patterns of biodiversity variation has had a high priority in recent overviews of the issue (for instance, in the "Insight" published in *Nature* in May 2000). Ecological change, whether driven by natural- or human-induced dynamics (which, furthermore, mingle together), makes it imperative to include a temporal aspect in all assessment schemes. Margules et al. (1994) demonstrated the significance of distributional dynamics in an apparently simple system of limestone pavements in Yorkshire; the dynamic aspect of biodiversity preservation is emphasized, for instance, in Haila and Kouki (1994) and in the essays in Perrings et al. (1995) and Mace et al. (1998).

The search for indicators and surrogates aims at formulating specified research tasks. This research framing has roots reaching back to the early twentieth century. Plant ecologists, in particular, were interested in the variation in species richness as a function of area. With the development of efficient sampling methods, zoologists became interested in patterns of variation in species richness as a function of sample size (Fisher et al. 1943; Williams 1964). Island Biogeography, which I mentioned in the previous section, is one outgrowth from this tradition. The relationship between local and regional processes in shaping species richness was recognized as a critical research problem within this research tradition, and a whole range of causal hypotheses have been suggested. The essays in Diamond and Case (1986) and Ricklefs and Schluter (1993) offer a rich overview of this question.

2.4.2 *Ontological Commitments*

The choice of indicators and surrogates for describing variation in biodiversity cannot be based on purely technical decisions, however. The indicators and surrogates also have to be justified theoretically, albeit perhaps indirectly, by suggesting some good causal reasons to regard a particular surrogate as reliable. Theoretical justifications usually lean ultimately on the importance of species in biology.

E. O. Wilson has defended a strong view on the role of species in biology. In his popular monograph he argues as follows:

> Western science is built on the obsessive and hitherto successful search for atomic units, with which abstract laws and principles can be derived. Scientific knowledge is written in the vocabulary of atoms, subatomic particles, molecules, organisms, ecosystems, and many other units, including species.... So, the species concept is crucial to the study of biology. It is the grail of systematic biology. (1992, 35–6)

We need to pause again and consider the argument in some detail. Wilson's argument is built on what Western science in general is supposed to have achieved, and how. Hence, it is clearly logically different from an argument built on the *biological significance* of species. Recently Wilson expressed the same idea using stronger words (Wilson 2003, 77). He presented a plea for an "encyclopedia of life." The core of the plea is a suggestion to construct an encyclopedia, which would include an electronic page for each species of organisms on Earth. This would, Wilson suggests, have "heuristic power for biology as a whole. As the census of species on earth comes ever closer to completion, and as their individual pages fill out to address all levels of biological organization from gene to ecosystem, new classes of phenomena will come to light at an accelerating rate."

Quite frankly, the last sentence in Wilson's phrasing is difficult to follow. It stipulates that a complete encyclopedia of species would allow total derivability both toward genes, downward in the biological hierarchy, and toward ecosystems, upward in the hierarchy, *simultaneously*. It is hard to grasp how this could be achieved. There is ample literature available, which shows that such unambiguous interlevel derivations across the biological organizational hierarchy simply are not possible. Just to show that Wilson is serious with his strange suggestion, I take one more citation from his article: "Only with such encyclopedic knowledge can ecology mature as a science and acquire predictive power species by species, and from those, ecosystem by ecosystem."

One can surmise that Wilson's idea of an encyclopedia of life builds upon the Human Genome Project. Now that a practically complete map of the base sequences in the human genome has been around for some time, the limits of the analogy are readily visible. The "gene map" does not fulfill original promises that it would dramatically increase our understanding of human biology, the biology of diseases in particular. To demonstrate the difficulty, suffice it to cite David Baltimore's first assessment of the "gene map" (Baltimore 2001, 816): "Understanding what does give us our complexity – our enormous

behavioural repertoire, ability to produce conscious action, remarkable physical coordination (shared with other vertebrates), precisely tuned alterations in
response to external variations of the environment, learning, memory . . . need
I go on? – remains a challenge to the future."

Baltimore's assessment reflects the crucial refinement that molecular biology has brought to our understanding of genes, namely, that their effects are
contextual. Wilson's encyclopedia of life, if it ever materialized, would bring
similar problems into focus. Ecosystems are molded by interactive processes,
which take place on several temporal and spatial scales simultaneously. The
role of species in ecosystems is contextual analogously to the role of genes
in organisms. For instance, individuals of the same species end up having
different phenotypes and, hence, different ecological roles depending on the
developmental pathway that they went through in any particular environment.
Furthermore, the role of a particular species in any given ecosystem depends
on the array of other species present. There is simply no way of predicting
ecosystem dynamics from the characteristics of single species. Note, furthermore, that the biological species concept on which Wilson's views are based
is totally inadequate for the most important actors in ecosystems, namely,
micro-organisms (Margulis and Sagan 2002).

Wilson's plea leans primarily on his personal faith in the supremacy
of an atomistic research strategy. Remember the wording above: "the
obsessive . . . search for atomic units, with which laws and principles can
be derived." However, Wilson's atomistic plea is too simplistic. Western science cannot be reduced to one formula; rather, the sciences are disunited, and
different sciences have utilized in their history several alternative "styles of
reasoning," which cannot be directly translated into one another (Hacking
1996). Moreover, if one were hard pressed to identify a single successful explanatory strategy in the history of Western science, it would be *the search for
mechanisms*, not for fundamental units, laws, and principles (Hacking 1983;
Cartwright 1999).

The role of species in ecological systems is ultimately an empirical question. A successful research program on this problem cannot be built on a priori
restrictive views on what sort of entities species actually are. Species are not
unchanging entities. Margules and Pressey (2000, 248) give a pragmatic definition of species in an evolutionary context, as follows: "It has long been
argued that species should be treated as dynamic evolutionary units rather
than types." In an ecological context, dynamics needs even more emphasis.
One of the critical challenges of biodiversity research is to understand mechanisms, which produce and maintain heterogeneity in ecological systems. An
atomistic research strategy does not recognize this problem.

2.5 PROCESSES

2.5.1 Analytic Challenges

A basic aspect of biology is that all "entities" are temporary. They are maintained by particular reproductive processes. In the earliest stages of the history of life, there were no entities around at all, only reproductive cycles; Manfred Eigen and Peter Schuster captured this idea with their term "hypercycle" (Eigen et al. 1981). The maintenance of such cycles is a dynamic question. In the course of evolution, forms of life have increased in size, and become more complex and more stable, in a relative sense; this is one of the few secular trends in organic evolution (e.g., Bonner 1988; Morowicz 1992). It is always good to remember, however, that the world of multicellular macroscopic organisms really rests on a fabric of life maintained by the interactive reproductive dynamics of zillions of micro-organisms (Margulis and Sagan 2002).

In other words, a process perspective is necessary for understanding the existence of biological entities. This insight can be learned from Alfred Lotka's classic *Elements of Physical Biology* (Lotka 1956). As has often been noted, modern ecology has been largely divided into two branches – ecosystems ecology on the one hand, and population and community ecology on the other hand; or ecology of functions and ecology of entities, respectively (e.g., O'Neill et al. 1986). Textbooks and monographs on modern ecology seldom cross this division line.

Biological diversity cuts right through the division line between entities and functions: entities make up the patterns of heterogeneity and variability that we observe and call biodiversity, but reproductive processes are needed to maintain them. Elucidating this relationship presents two challenges for biodiversity research. One challenge is a problem for basic ecological research: What is the functional significance of biodiversity? Essays in Schulze and Mooney (1993) originally framed the problem conceptually. The question has inspired several experimental projects both in the field and in the laboratory; the results of the experiments were recently summarized in Kinzig et al. (2001). The other challenge is practical: Does the dynamic nature of ecological systems matter for conservation? The answer is yes. Pickett and Thompson (1978) were among the first to draw this conclusion, well before the slogan BioDiversity had been heard. The dynamic aspects of the biodiversity issue are emphasized in the essays in Mace et al. (1998); see also Haila (1999a) and Margules and Pressey (2000).

Also the choice of indicators and surrogates for surveying and monitoring biodiversity has a dynamic edge. The relative stability of the particular entities in question is a critical issue. Dynamic patterns cannot be preserved by assuming that they could be handled as independent entities, capable of being separated from the conditions of their reproduction. To take an absurdly obvious example, let's assume that someone, for some weird reason, would like to preserve tornadoes. Tornadoes are temporary phenomena, which occur quite predictably under the right conditions. In a sense a tornado is simultaneously a structure and a process, almost instantly dissipating enormous amounts of energy accumulated in the lower parts of the atmosphere (Bluestein 1999). What should be preserved to preserve a tornado? The answer is: the conditions producing tornadoes. It is simply impossible even to identify, let alone to "preserve," tornadoes as "entities" separate form the conditions bringing them about.

This example is not so far-fetched as it may seem. The same principle applies also to such "entities" which are in existence only through being necessary elements in ongoing biological processes. Consider genes. Recent progress in molecular biology has shown that cellular metabolism is, in fact, maintained by multilayered control mechanisms in which both genes and a whole range of polypeptides (enzymes) are continuously participating in cycles of continuous production, destructive use, and reproduction. The maintenance of cellular metabolism requires all these elements, but it only happens in the milieu provided by a living cell. No mixture of DNA strains and polypeptides alone can substitute for a living cell.

Moss (2003) provides the most efficient, philosophically informed demythologization of the concept of the gene to date. The bottom line is that genes alone do not do any job in the cell, or anywhere else for that matter. Genes do nothing at all except as participants in the metabolism of living cells. An implication is that genes cannot really be preserved in the shape of deep-frozen DNA strains. To function as genes, to *be* genes, bits of DNA must participate in the continuous functions of living cells. Of course, DNA molecules do have crucial functional roles in individual development and metabolism of organisms, but they are not intact entities as simplistic genetic models used to imply. Also, the view of the role of genes in evolution is being modified, as the following note by Hiroaki Kitano points out (Kitano 2002, 208): "It may be that functional circuits should be considered the units of evolution."[8]

In process terms, biodiversity is a pervasive characteristic of life. However, life on Earth is not a single, all-encompassing, dynamically unified system, but rather a complex of reproductive cycles. The general features of the reproductive dynamics of life are nowadays increasingly understood on the level of cells and organisms. In ecology, the situation is more moot because of the "bifurcation" into two separate research traditions discussed above.

2.5.2 Practical Challenges

An ultimate paradox of the centrality of reproductive process is that for preservation alone we do not, ultimately, need to know the "entities" that participate in the process. As long as conditions are right, the functions will go on and take care of the entities; in other words, if the conditions are right, biodiversity will take care of itself (Haila 1995; Holling 1995). For instance, when an organism is alive, we know for sure that the important molecules are in place, continuously reproduced by the metabolic processes of the organism. If the organism belongs to a large, panmictic population, we know that both the genes and the cellular machinery will be reliably transmitted from one generation to the next. This knowledge doesn't become a bit more reliable if we get a complete catalogue of the genes and other molecules and cell organelles that are transmitted.

However, on the macroscopic level, the practical task of biodiversity preservation is not quite that simple. Ecosystems, in which populations lead their lives, are not alive or dead in the same sense as organisms. This means that *assessing the dynamic relevance* of the indicators and surrogates used in biodiversity research is a critical task. Endangered species, which are the ultimate focus of the extinction concern, would be perfect indicators of themselves. The first priority would be to know precisely which ones they are, and where they are. For obvious reasons, this is seldom feasible. One possibility is to use so-called umbrella species as surrogates, in other words, species that are not quite as rare as the endangered ones but share with them some critical ecological characteristics or needs (for the concept and examples, see Wilcove 1994 and Martikainen et al. 1998). More indirectly, habitats and environmental types hosting many rare species also offer potential as surrogates. An efficient possibility is to map the environmental space that is on offer for various species in a particular region; efficient statistical procedures are available, provided adequate surveys of the spatial distribution of relevant habitat elements are available. Pioneering research was conducted on this particular problem in Australia in the pre-BioDiversity era (Austin et al. 1984); for later developments see Margules (1989) and Margules and Austin (1991).

In any case, the dynamic processes that maintain critical population and ecosystem characteristics must be identified and respected. Rapid turnover belongs to the nature of biological processes. To clarify the case, I once again cite an extremist argument presented by Paul Ehrlich (1995, 214). Ehrlich's argument is that as species in nature are divided into more or less distinct populations, and as these local populations are "disappearing at a high rate," the extinction crisis is actually much more serious than mere species lists let us understand. He articulates his extremism on this issue as follows: "If population extinctions reduced all remaining species to single minimum viable populations, no further species extinctions would have happened. Nonetheless, an extinction catastrophe would have taken place that might well, through interruption of ecosystem services, cause the demise of humanity as well" (references omitted).

Once again, we need to explore the argument. First, it is established knowledge in ecology, of course, that natural populations are divided into more or less isolated subpopulations; the basic text is Andrewartha and Birch (1954). The turnover rate of local populations may be very high in perfectly natural conditions, for instance, under disturbance regimes driven by wildfires, storms, floods, droughts, pest outbreaks, and so on. If we include processes on longer temporal scales, we get to the dynamics of climatic fluctuations and continental glaciations. So, what to say about those parts of the northern continents which were devoid of higher organisms only 10,000 years ago? The colonization of these regions after the continental ice was gone was no smooth, monotonous process. The latest abrupt transformation in ecosystem structure in Finland and Scandinavia, for instance, took place with the invasion of spruce some 4,000 years ago. A whole range of "local populations" certainly went under, replaced by aggressive invaders following the spruce. Figures estimating these local population turnover events would be very high indeed, albeit impossible to estimate realistically. Fortunately, they are irrelevant.

The point is that local population turnover is an integral part of population dynamics, similar to genetic shuffling and reshuffling in sexually reproducing populations. It wouldn't make sense to say that genes go "locally extinct" whenever an individual dies. The case of local populations is similar.[9] A presumption that such turnover is accompanied by "loss of ecosystem services" has to be grounded in specified hypotheses on what is the mechanism that could cause such a loss.

This connects us to the other major fault in Ehrlich's gloomy prediction: it lacks a credible mechanism. No conceivable mechanism exists through which human activity could reduce "all remaining species to single minimum viable populations," as Ehrlich writes. This matter is serious. Conservationists, when uttering gloomy predictions supposedly drawing upon science, should be sure that credible mechanisms exist that might actually cause the harm they are warning about. Of course, human actions do often reduce diversity in ecosystems, but always through specific processes; a good overview of different ways this happens is described in Lawler et al. (2001). It is highly unlikely that all the mechanisms they suggest would be acting simultaneously in any specific context, but a specified list is a good start.[10]

2.6   HUMAN SUBSISTENCE: FROM MALTHUSIAN CLOSURE
TO SOCIOECOLOGICAL ANALYSIS

As discussed previously, human population growth is commonly named as the cause of the biodiversity crisis. It is easy indeed to find generalizing statements to this effect. These assessments have a double basis: First, an assumption that the relationship between human subsistence and the productivity of other ecological processes is a zero-sum game, that is, the relationship is fruitfully addressed within a Malthusian closure. The second element is the fallacy of global averages, which I discussed earlier.

These, of course, belong together. A famous example of zero-sum logic is the rule $I = P*A*T$: *Impact = Population* $\times$ *Affluence* $\times$ *Technology* (Ehrlich and Ehrlich 1990, 58). The formula hides three basic assumptions. First of all, each one of the terms on the right-hand side is assumed to be quantifiable in a straightforward manner. For this to be true, each of them should be composed of indivisible basic elements, which can be added up to get the total value. Second, the factors connect together directly without interaction terms; for instance, increasing affluence cannot facilitate increase of human skills or development of new technologies such that the impact on the environment declines. Third, Earth is assumed to be a unified and uniform whole: the impact is the same irrespective of whether the person in question lives in Greenland or in Amazonia.

None of the assumptions is true. In fact, the formula $I = P*A*T$ is nonsense. Nevertheless, amazingly numerous and blatant examples of Malthusian logic in biodiversity discourse could be named; but let this be, it would not be worthwhile.

It is much more important to emphasize that Malthusian ruminations form only one, albeit visible, strand in discussions about biodiversity and human subsistence. Assessing this relationship poses serious questions for empirical research, as ecologists have realized many times over. Essays in several recent collections demonstrate this point. Ecological effects of human activities are assessed in McDonnell and Pickett (1993). Traditional subsistence systems are evaluated in Berkes and Folke (1998). C. S. Holling with his colleagues has published a series of works on shared dynamics of human productive systems and ecosystem processes; these are summarized in Gunderson and Holling (2002). Co-evolution of human sustenance and ecological processes is an attractive metaphor, which Norgaard (1994) has developed within an economic-anthropological framework.

The preservation of endangered species can be framed as a co-evolutionary challenge as well. Once species preservation succeeds, the populations of the target species begin to increase. This is a universal consequence of successful conservation, whether we are talking about wolves, bears, and elephants, or whales and seals. So, at some stage humans living next to the increasing populations of such species have to decide how to get along with them. Local conflicts between seal populations and coastal fisheries in the Baltic or the Norwegian Sea, or between elephants and cattle herding inside and outside of African nature parks, are well known. It is simply a necessity that local human subsistence must adapt to the presence of those species. But the reverse is also true: the preserved species have to behave when they are close to human habitation. For instance, predators such as wolves and bears, who get used to eating garbage at holiday resorts, will turn to eating hikers when the need and the opportunity arise.[11]

A critical theoretical challenge in this endeavor is scale correspondence between ecological processes and human practices (Saunders and Briggs 2002). Serious attention needs to be paid to habitat degradation, which is certainly a very common feature in productive landscapes, but this requires specification of critical scales. Local people can be motivated to improve their own environments when given the chance.

In fact, the appreciation of nature *everywhere*, including the habitats molded and dominated by humans, is a key challenge for us to build up harmonic relationships between human culture and the rest of nature. At this point, the nature–culture dualism, which is strengthened by mystification of "wilderness" supposedly untouched by human beings, becomes a particular burden (see Haila 2000). As Wes Jackson (1991, 51) writes: "In other words, Harlem and East Saint Louis and Iowa and Kansas and the rest of the world where wilderness has been destroyed have to be loved by enough of us, or wilderness is doomed."

2.7 CONCLUDING REMARKS: WHICH WAY NEXT?

The preceding survey of the framing of biodiversity invites the following summarizing points. First of all, the survey has revealed a tension between political and conceptual framing of the issue. The political framing has been dominated ever since the Washington forum by a strong urge to use global terms and arguments, that is, to make the issue as "big" as possible. This does not, however, suggest any specified research. The political and conceptual framings do not meet.

A remarkable feature of biodiversity among environmental concerns is the widely shared consensus on the importance and "big" size of the issue. In fact, in a cynical mood one might suspect that this is a self-fulfillment of a slogan in Takacs (1996, 99): "If biodiversity is a much more complex and dynamic focus for conservation efforts than endangered species, it likewise offers a much more complex and dynamic role for biologists in society at large." So, everybody who doubts a single argument expressed by a biodiversity biologist offends the professional prestige of the whole crowd and has to be silenced?

Recent public debates around the issue suggest that such a suspicion has some truth in it. The reception of Bjørn Lomborg's (2001) recent book criticizing environmentalist arguments offers an example. Lomborg also criticizes in his book the estimates of extinction rates commonly used in biodiversity discussions. As a consequence, many active proponents of the biodiversity issue have refused to consider Lomborg's arguments at all. Paul Ehrlich, for instance, finished his review of the book with the following sentences: "Yes, environmental scientists make mistakes, and Lomborg reports some of them. But useful debate occurs only among those who have demonstrated that they understand the situations about which they are writing" (Ehrlich 2002).

I am no particular fan of Lomborg's book (in fact, I find his boy-scout optimism rather sickening), but he deserves the right to be taken seriously. This is the case especially for the paradoxical reason that Lomborg's basic mistake, a heavy reliance on global statistics, is in fact a mirror image of the global arguments used by many environmentalists. Ehrlich's comment cited above thus invites a question: What would be a way to ensure that the participants in the debate actually "understand the situations"? Should we, perhaps, use the acceptance of Ehrlich's own predictions as a critical test? I am afraid that not many participants would be left in the discussion if everybody had to accept, for instance, that because of the extinction crisis, "civilization will disappear some time before the end of the next century" (Ehrlich 1988).

This might be funny if the issue were not so serious. In fact, the Lomborg controversy only repeats an earlier nondebate between Julian Simon and Paul Ehrlich. The arguments Lomborg gives concerning the unreliability of the current estimates of extinction rates are basically identical with Simon's (Simon 1996). Ehrlich's response to Lomborg is basically similar to his response to Simon's earlier arguments. Simon and Lomborg are the flip side of the "crying crisis" of early environmentalism, and the continuous extremism of Paul Ehrlich is the flip side of Simon and Lomborg.

It seems that these extremist views put forth by both sides of the controversy have supported each other, and this mutual entrainment has produced a new problem space within which the biodiversity crisis is evaluated. The space is delineated by a contrast between *global optimism versus global pessimism.* This is a false and counterproductive contrast. It leaves no space for realistic assessment of specific problems.

We are justified to conclude that the original framing of biodiversity as an extinction crisis has had a continuous influence on the understanding of the issue. One specific negative effect is that such research that needs to be done is not done. In particular, too little is known of the specific ecological relationships between preserves and the surrounding human-modified areas. The absence of Lugo's (1988) paper – published in the proceedings of the Washington forum – from later assessment is symptomatic of this problem. There is a dearth of data on how various organisms actually survive in human-modified environments. The adherence to the Island Biogeography model of extinction assessment does not allow this question to be asked at all (Haila 1990, 2002a). Simple and concrete questions to ask are, for instance, What are the minimum requirements of organisms in managed areas? and How good are the prospects of securing such requirements under different management regimes? (Haila et al. 1989, 1994).

The normative aspect of the biodiversity issue rests implicitly on a distinction between normal and abnormal states of ecological systems (see Haila 1999a; Sarkar 2002). A basic part of this argumentation is the claim that humans excessively dominate Earth's ecosystems. Wilson is explicit in this regard:

> By every conceivable measure, humanity is ecologically abnormal. Our species appropriates between 20 and 40 percent of the solar energy captured in organic material by land plants. There is no way that we can draw upon the resources of the planet to such a degree without drastically reducing the state of most other species. (1992, 260)

This is a complicated issue. Wilson, in the preceding quotation, uses ecology as a simple quantitative standard. Leaving aside the fact that estimating the human appropriation of solar energy is very problematic, the view is problematic because by the same reasoning, any dominant life-form would be "abnormal" in its living environment. For instance, the invasion of spruce brought about a dramatic change in northwest European forests, and most probably a decline in ecological diversity of forests. Should we think that the spruce is ecologically abnormal?

This wouldn't make sense. The crux of the matter is that ecology can be used as a measure for evaluating human "normality" versus "abnormality" only concerning concrete matters such as specified historical situations and ecological settings, and only if *humans themselves are included in the same equation.*

The weight of the extinction argument throughout the career of the biodiversity concern attests to the path-dependence of the process. Early warnings remain alive, and interfere with analytic framings of the issue by their sheer weight. I would suspect that this also has a national dimension: the political event of the Washington Forum on BioDiversity in 1986 could hardly have taken place in any other country than the United States, with its strong tradition of political advocacy.

This brings up a paradox of advocacy science: the significance of alarm calls cannot be evaluated by research results alone. That the public at large has been more or less persuaded about the reality of an extinction crisis has also stimulated shifting perceptions of the human ecological predicament. Warnings always have effects, although these are seldom immediately visible; indeed, in "preventive fights," the lack of visible effects may be the main effect – but this can never be proven. A similar setting is true of other areas, too; for instance, the claim "the anti-war movement was unnecessary because no war broke out" is not logically valid.

This point is historically plausible. There is no need to feel guilty about the strong phrases used in the 1980s about extinction threats. Space had to be conquered for the issue in the public domain. However, the situation has changed. In particular, the dynamics of the ecological basis of human subsistence needs to be addressed far more specifically than has been the prevailing habit. This is the dimension of the biodiversity issue that has suffered most from sloppy and global argumentation. On socioecological dynamics, papers with amazingly low analytic standards are being published on prominent forums. If you doubt this, take a look at Dobson et al. 1997 (particularly their section "Modelling habitat conversion," 516–17), and Chapin III et al. 2000; particularly their Fig. 1). These papers rely on such broad categories that the conclusions are analytically worthless. Fortunately, however, positive examples abound, too.

It seems that empirical research on biodiversity is divided into several semi-independent genres. In this survey I have referred to (1) estimating extinction rates (most intimately connected to the Malthusian framing); (2) assessing geographical patterns in biodiversity variation; (3) developing methods of surveying and monitoring biodiversity; (4) evaluating the functional significance of biodiversity in ecosystems; and (5) assessing the relationships between bio-diversity and human subsistence practices. Several important fields of study such as conservation genetics are missing from this list, which is not meant to be comprehensive. The point is, rather, to suggest that biodiversity has triggered and inspired several rapidly developing research fields, which do not form a conceptually coherent discipline. Each one of the traditions draws upon flourishing research in nearby fields, as well as upon older research traditions. It is here that successful "mangling" (Pickering 1995) occurs. We have no need to be worried even though no logically coherent single "theory about biodiversity" will show up.

Finally, my survey of the issue gives rise to a normative conclusion, too. I think a major goal is to bring the empirically strong research strains into fruitful contact with a realistic understanding of the political process of con-servation. Scaling is a critical issue in this context, particularly understanding the multiplicity and specificity of critical scales of dynamic change on nature on the one hand, and human subsistence practices on the other hand. Political processes have their own characteristic scales that are equally real and press-ing as the scales in more tangible parts of nature (Meadowcroft 2002). There are no a priori reasons to expect that the relevant scales of political processes and ecological processes would match each other.

Getting empirical biological research and creative political thinking and action into fruitful interaction with each other presents the greatest challenge for a process view on biodiversity.

NOTES

1. Takacs's book is, in essence, a hagiography of the concept of "biodiversity"; his arguments are often irritatingly vague. However, the book is based on interviews, which seem to be reliably recorded. So, the book can be taken as reflecting the self-understanding of those involved in the early promulgation of biodiversity – albeit perhaps with a grain of salt.
2. I consider this very strange: at the very least, people who do not accept Lugo's points should present their arguments publicly, particularly while Lugo was con-cerned about tropical forests. Heywood and Stuart (1992) and Whitmore (1997) make similar points concerning our poor knowledge about relationships between fragments and their surroundings in tropical forests.
3. The figure is about 1 percent of the total number of bird species, which is some 9,500; the figures vary a bit in different sources, but this is inconsequential for the argument.
4. Let's make two clarifications so that my argument is not blatantly misunderstood. First, any secular trend over the last centuries correlates positively (or negatively, as the case may be) with human population growth. This is no demonstration of causation. Second, humanity has certainly become a geological force on Earth, as

George Perkins Marsh concluded one and a half centuries ago (Marsh 1965). This doesn't, however, imply a Malthusian closure on the global scale. In many specific local and regional contexts, on the other hand, a Malthusian relationship between human population and particular resources is very much true today; the food crisis in sub-Saharan Africa offers examples.

5. An article by Crutzen and Andreae (1990) on the burning of biomass in the tropics is as good an example as any; the authors do not even raise the question whether any means for disposing of waste biomass would be available other than burning.

6. At this point, before we rush to object that nobody would take the surgery metaphor literally, it is sobering to remember Garrett Hardin's (1974) "life-boat ethics."

7. The terms indicator and surrogate are not synonymous, although their meanings overlap. Surrogates are more distanced from the substance of biodiversity, as it were; for instance, habitat area can function as a surrogate in a particular context because of the general positive correlation between area and species richness, but area is not a valid indicator.

8. Just pick up a population genetics/evolutionary biology textbook of the 1970s or 1980s to get a feel for the enormity of this shift. Nowadays, almost every issue of *Nature* or *Science* includes articles that testify to the revolutionary change that has taken place in our understanding of what genes do. To clarify the point further, consider the number of genes; Aebersold (2003) writes as follows: "Although the number of genes analysed to date is relatively small . . . the number of encoded proteins is enormous, as the same gene can generate multiple protein products that differ as a result of combinatorial splicing, processing and modification." This is in stark contrast with an authorized textbook formula from the 1970s (Dobzhansky et al. 1977, 28): "one gene – one polypeptide."

9. Literature demonstrating the highly dynamic character of local ecological assemblages in all major biomes of Earth is vast; a good place to start is Gunderson and Holling (2002).

10. Another example of an unspecified gloomy prediction from a prestigious essay: Norman Myers (1997, 132) writes: "There is likely to be a significant disruption of certain basic processes of evolution." However, he does not give/cannot give any specification on what the mechanisms of this disruption actually are.

11. Community-based conservation is a potential way to achieve a co-evolutionary path, but it is controversial among conservationists; see the exchange between David Western and John Terborgh in *New York Review of Books*, August 12, 1999. Biologists studying endangered species may be sadly oblivious to the needs of local people: Thompson (2002) tells a story on this about elephants in the Amboseli Nature Park, Kenya.

BIBLIOGRAPHY

Aebersold, R. 2003. Constellations in a Cellular Universe. *Nature* 422:115–16.

Andrewartha, H. G. and L. C. Birch. 1954. *The Distribution and Abundance of Animals.* Chicago: University of Chicago Press.

Austin, M. P., R. B. Cunningham, and P. M. Fleming. 1984. New Approaches to Direct Gradient Analysis Using Environmental Scalars and Statistical Curve-Fitting Procedures. *Vegetatio* 55:11–27.

Baltimore, D. 2001. Our Genome Unveiled. *Nature* 409:814–16.

Berkes, F. and C. Folke, eds. 1998. *Linking Social and Ecological Systems. Management Practices and Social Mechanisms for Building Resilience.* Cambridge: Cambridge University Press.

Bluestein, H. B. 1999. *Tornado Alley. Monster Storms of the Great Plains.* New York: Oxford University Press.

Bonner, J. T. 1988. *The Evolution of Complexity by Means of Natural Selection.* Princeton, NJ: Princeton University Press.

Cartwright, N. 1999. *The Dappled World. A Study of the Boundaries of Science.* Cambridge: Cambridge University Press.

Chapin III, F. S., et al. 2000. Consequences of Changing Biodiversity. *Nature* 405:234–42.

Crutzen, P. J. and M. O. Andreae. 1990. Biomass Burning in the Tropics: Impacts on Atmospheric Chemistry and Biogeochemical Cycles. *Science* 250:1669–78.

Diamond, J. and T. J. Case, eds. 1986. *Community Ecology.* New York: Harper & Row.

Dobson, A. B., A. D. Bradshaw, and A. J. M. Baker. 1997. Hopes for the Future: Restoration Ecology and Conservation Biology. *Science* 277:515–22.

Dobzhansky, T., F. J. Ayala, G. L. Stebbins, and J. W. Valentine. 1977. *Evolution.* San Francisco: W. H. Freeman.

Dyke, C. 1981. *Philosophy of Economics.* Englewood Cliffs, NJ: Prentice-Hall.

    1988. *The Evolutionary Dynamics of Complex Systems. A Study in Biosocial Complexity.* Oxford: Oxford University Press.

    1992. From Entropy to Economy: A Thorny Path. *Advances in Human Ecology* 1:149–76.

    1993. Extralogical Excavations. Philosophy in the Age of Shovelry. In *Foucault and the Critique of Institutions*, J. Caputo and M. Yount, eds. Philadelphia: Pennsylvania State University Press.

    1994. The World Around Us and How We Make It: Human Ecology as Human Artefact. *Advances in Human Ecology* 3:1–22.

Ehrlich, P. R. 1988. The Loss of Diversity: Causes and Consequences. In *Biodiversity*, E. O. Wilson and F. M. Peter, eds. Washington, DC: National Academy Press.

    1995. The Scale of the Human Enterprise and Biodiversity Loss. In *Extinction Rates*, J. H. Lawton and R. M. May, eds. Oxford: Oxford University Press.

    2002. The Brownlash Rides Again. *Trends in Ecology and Evolution* 17:51.

Ehrlich, P. R. and A. H. Ehrlich. 1981. *Extinction: The Causes and Consequences of the Disappearance of Species.* New York: Ballantine Books.

    1990. *The Population Explosion.* New York: Simon & Schuster.

Eigen, M., W. Gardiner, P. Schuster, and R. Winkler-Oswatitsch. 1981. The Origin of Genetic Information. *Scientific American* 244:78–94.

Einarsson, N. 1993. All Animals Are Equal but Some Are Cetaceans: Conservation and Culture Conflict. In *Environmentalism. The View from Anthropology*, K. Milton, ed. London: Routledge.

Fisher, R. A., A. S. Corbet and C. B. Williams. 1943. The Relation between the Number of Individuals and Number of Species in a Random Sample of an Animal Population. *Journal of Animal Ecology* 12:42–58.

Garfinkel, A. 1981. *Forms of Explanation. Rethinking the Questions in Social Theory.* New Haven, CT: Yale University Press.

Gunderson, L. H. and C. S. Holling, eds. 2002. *Panarchy. Understanding Transforma-tions in Human and Natural Systems*. Washington, DC: Island Press.

Hacking, I. 1983. *Representing and Intervening. Introductory Topics in the Philosophy of Natural Science*. Cambridge: Cambridge University Press.

——— 1992. The Self-Vindication of the Laboratory Sciences. In *Science as Practice and Culture*, A. Pickering, ed. Chicago: University of Chicago Press.

——— 1996. The Disunities of the Sciences. In *The Disunity of Science. Boundaries, Con-texts, and Power*, P. Galison and D. J. Stump, eds. Stanford, CA: Stanford University Press.

——— 1999. *The Social Construction of What?* Cambridge, MA: Harvard University Press.

Haila, Y. 1990. Toward an Ecological Definition of an Island: A Northwest European Perspective. *Journal of Biogeography* 17:561–8.

——— 1994. Preserving Ecological Diversity in Boreal Forests: Ecological Background, Research, and Management. *Annales Zoologici Fennici* 31:203–17.

——— 1995. Natural Dynamics as a Model for Management: Is the Analogue Practicable? In *Northern Wilderness Areas: Ecology, Sustainability, Values*, A-L. Sippola, P. Alaraudanjoki, B. Forbes, and V. Hallikainen, eds. Rovaniemi: Arctic Centre Publications 7.

——— 1998. Political Undercurrents of Modern Ecology. *Science as Culture* 7:465–91.

——— 1999a. Biodiversity and the Divide between Culture and Nature. *Biodiversity & Conservation* 8:165–81.

——— 1999b. Islands and Fragments. In *Maintaining Biodiversity in Forest Ecosystems*, M. L. J. Hunter, ed. Cambridge: Cambridge University Press.

——— 1999c. Socioecologies. *Ecography* 22:337–48.

——— 2000. Beyond the Nature–Culture Dualism. *Biology & Philosophy* 15:155–75.

——— 2002a. A Conceptual Genealogy of Fragmentation Research: From Island Biogeog-raphy to Landscape Ecology. *Ecological Applications* 12:321–34.

——— 2002b. Scaling Environmental Issues: Problems and Paradoxes. *Landscape and Urban Planning* 61:59–69.

Haila, Y., I. K. Hanski, J. Niemelä, P. Punttila, S. Raivio, and H. Tukia. 1994. Forestry and the Boreal Fauna: Matching Management with Natural Forest Dynamics. *Annales Zoologici Fennici* 31:187–202.

Haila, Y., I. K. Hanski, and S. Raivio. 1989. Methodology for Studying Minimum Habitat Requirements of Forest Birds. *Annales Zoologici Fennici* 26:173–80.

Haila, Y. and O. Järvinen. 1990. Northern Conifer Forests and Their Bird Species As-semblages. In *Biogeography and Evolution of Forest Bird Communities*, A. Keast, ed. The Hague: SPB Academic Publishers.

Haila, Y. and J. Kouki. 1994. The Phenomenon of Biodiversity in Conservation Biology. *Annales Zoologici Fennici* 31:5–18.

Haila, Y. and R. Levins. 1992. *Humanity and Nature. Ecology, Science and Society*. London: Pluto Press.

Hannigan, J. A. 1995. *Environmental Sociology. A Social Constructionist Perspective*. London: Routledge.

Hardin, G. 1974. Living in a Lifeboat. *BioScience* 24:561–7.

Heywood, V. H. and S. N. Stuart. 1992. Species Extinctions in Tropical Forests. In *Tropical Deforestation and Species Extinction*, T. C. Whitmore and J. A. Sayer, eds. London: Chapman and Hall.

Holling, C. S. 1995. What Barriers? What Bridges? In *Barriers & Bridges to the Renewal of Ecosystems and Institutions*, L. H. Gunderson, C. S. Holling, and S. S. Light, eds. New York: Columbia University Press.

Hopper, S., J. Chappill, M. Harvey, and A. George, eds. 1996. *Gondwanan Heritage. Past, Present and Future of Western Australian Biota.* Chipping Norton, NSW: Surrey Beatty & Sons.

Jablonski, D. 1995. Extinctions in the Fossil Record. In *Extinction Rates*, J. H. Lawton and R. M. May, eds. Oxford: Oxford University Press.

Jackson, W. 1991. Nature as a Measure for a Sustainable Agriculture. In *Ecology, Economics, Ethics. The Broken Circle*, F. H. Bormann and S. R. Kellert, eds. New Haven, CT: Yale University Press.

Jacobs, J. 2000. *The Nature of Economies.* New York: The Modern Library.

Järvinen, O. and R. A. Väisänen, 1979. Changes in Bird Populations as Criteria of Environmental Changes. *Holarctic Ecology* 2:75–80.

Kinzig, A. P., S. W. Pacala, and D. Tilman, eds. 2001. *The Functional Consequences of Biodiversity. Empirical Progress and Theoretical Extensions.* Princeton, NJ: Princeton University Press.

Kitano, H. 2002. Computational Systems Biology. *Nature* 420:206–10.

Lawler, S. P., J. J. Armesto, and P. Kareiva. 2001. How Relevant to Conservation Are Studies Linking Biodiversity and Ecosystem Function? In *The Functional Consequences of Biodiversity. Empirical Progress and Theoretical Extensions*, A. P. Kinzig, S. W. Pacala, and D. Tilman, eds. Princeton, NJ: Princeton University Press.

Lomborg, B. 2001. *The Skeptical Environmentalist. Measuring the Real State of the World.* Cambridge: Cambridge University Press.

Lotka, A. J. 1956. *Elements of Mathematical Biology* (Original: *Elements of Physical Biology*, 1924). New York: Dover.

Lovejoy, T. E. 1997. Biodiversity: What Is It? In *Biodiversity II. Understanding and Protecting Our Biological Resources*, M. L. Reaka-Kudla, D. E. Wilson, and E. O. Wilson, eds. Washington, DC: Joseph Henry Press.

Lugo, A. E. 1988. Estimating Reductions in the Diversity of Tropical Forest Species. In *Biodiversity*, E. O. Wilson and F. M. Peter, eds. Washington, DC: National Academy Press.

MacArthur, R. H. and E. O. Wilson. 1967. *The Theory of Island Biogeography.* Princeton, NJ: Princeton University Press.

Mace, G. M., A. Balmford, and J. R. Ginsberg, eds. 1998. *Conservation in a Changing World.* Cambridge: Cambridge University Press.

Margules, C. R. 1989. Introduction to Some Australian Development in Conservation Evaluation. *Biological Conservation* 50:1–11.

Margules, C. R. and M. P. Austin, eds. 1991. *Nature Conservation: Cost Effective Biological Surveys and Data Analysis.* Australia: Csiro.

Margules, C. R. and M. P. Austin. 1995. Biological Models for Monitoring Species Decline: The Construction and Use of Databases. In *Extinction Rates*, J. H. Lawton and R. M. May, eds. Oxford: Oxford University Press.

Margules, C. R., A. O. Nicholls, and M. B. Usher. 1994. Apparent Species Turnover, Probability of Extinction and the Selection of Nature Reserves: A Case Study of the Ingleborough Limestone Pavements. *Conservation Biology* 8:398–409.

Margules, C. R. and R. L. Pressey. 2000. Systematic Conservation Planning. *Nature* 405:243–53.

Margulis, L. and D. Sagan. 2002. *Acquiring Genomes. A Theory of the Origins of Species.* New York: Basic Books.

Marsh, G. P. 1965. *Man and Nature. Or, Physical Geography as Modified by Human Action* (David Lowenthal, ed.; originally published in 1864). Cambridge, MA: Belknap Press.

Martikainen, P., L. Kaila, and Y. Haila. 1998. Threatened Beetles in White-Backed Woodpecker Habitats. *Conservation Biology* 12:293–301.

McDonnell, M. J. and S. T. A. Pickett, eds. 1993. *Humans as Components of Ecosystems. The Ecology of Subtle Human Effects and Populated Areas.* New York: Springer.

Meadowcroft, J. 2002. Politics and Scale: Some Implications for Environmental Governance. *Landscape and Urban Planning* 61:169–79.

Mokyr, J. 1990. *The Lever of Riches. Technological Creativity and Economic Progress.* Oxford: Oxford University Press.

Morowitz, H. J. 1992. *Beginnings of Cellular Life. Metabolism Recapitulates Biogenesis.* New Haven, CT: Yale University Press.

Moss, L. 2003. *What Genes Can't Do.* Cambridge, MA: MIT Press.

Myers, N. 1979. *The Sinking Ark.* Oxford: Pergamon Press.

——— 1997. The Rich Diversity of Biodiversity Issues. In *Biodiversity II. Understanding and Protecting Our Biological Resources*, M. L. Reaka-Kudla, D. E. Wilson, and E. O. Wilson, eds. Washington, DC: Joseph Henry Press.

Norgaard, R. B. 1994. *Development Betrayed. The End of Progress and a Coevolutionary Revisioning of the Future.* London: Routledge.

O'Neill, R. V., D. L. DeAngelis, J. B. Waide, and T. F. H. Allen. 1986. *A Hierarchical Concept of Ecosystems.* Princeton, NJ: Princeton University Press.

Perrings, C., K-G. Mäler, C. Folke, C. S. Holling, and B-O. Jansson, eds. 1995. *Biodiversity Loss. Economic and Ecological Issues.* Cambridge: Cambridge University Press.

Pickering, A. 1995. *The Mangle of Practice. Time, Agency, and Science.* Chicago: University of Chicago Press.

Pickett, S. T. A. and J. N. Thompson. 1978. Patch Dynamics and the Design of Nature Reserves. *Biological Conservation* 13:27–37.

Preston, F. W. 1962. The Canonical Distribution of Commonness and Rarity. *Ecology* 43:185–215, 410–32.

Raup, D. M. 1988. Diversity Crisis in the Geological Past. In *Biodiversity*, E. O. Wilson and F. M. Peter, eds. Washington, DC: National Academy Press.

Reaka-Kudla, M. L., D. E. Wilson, and E. O. Wilson, eds. 1997. *Biodiversity II. Understanding and Protecting Our Biological Resources.* Washington, DC: Joseph Henry Press.

Ricklefs, R. E. and D. Schluter, eds. 1993. *Species Diversity in Ecological Communities. Historical and Geographical Perspectives.* Chicago: University of Chicago Press.

Sarkar, S. 2002. Defining "Biodiversity"; Assessing Biodiversity. *The Monist* 85:131–55.

Saunders, D. A. and S. V. Briggs. 2002. Nature Grows in Straight Lines – or Does She? What Are the Consequences of the Mismatch between Human-Imposed Linear

Boundaries and Ecosystem Boundaries? An Australian Example. *Landscape and Urban Planning* 61:71–82.

Schulze, E.-D. and H. A. Mooney, eds. 1993. *Biodiversity and Ecosystem Function*. New York: Springer Verlag.

Simberloff, D. 1997. Biogeographic Approaches and the New Conservation Biology. In *The Ecological Basis of Conservation. Heterogeneity, Ecosystems, and Biodiversity*, S. T. A. Pickett, R. S. Ostfeld, M. Shachak, and G. E. Likens, eds. New York: Chapman and Hall.

Simon, J. L. 1996. *The Ultimate Resource 2*. Princeton, NJ: Princeton University Press.

Soulé, M. E. 1985. What Is Conservation Biology? *BioScience* 11:727–34.

Steadman, D. M. 1997. Human-Caused Extinction of Birds. In *Biodiversity II. Understanding and Protecting Our Biological Resources*, M. L. Reaka-Kudla, D. E. Wilson, and E. O. Wilson, eds. Washington, DC: Joseph Henry Press.

Takacs, D. 1996. *The Idea of Biodiversity. Philosophies of Paradise*. Baltimore: Johns Hopkins University Press.

Thompson, C. 2002. When Elephants Stand for Competing Philosophies of Nature: Amboseli National Park, Kenya. In *Complexities. Social Studies of Knowledge Practices*, J. Law and A. Mol, eds. Durham, NC: Duke University Press.

Usher, M. B., ed. 1986. *Wildlife Conservation Evaluation*. London: Chapman and Hall.

Whitmore, T. C. 1997. Tropical Forest Disturbance, Disappearance, and Species Loss. In *Tropical Forest Remnants. Ecology, Management, and Conservation of Fragmented Communities*, W. F. Laurance and R. O. J. Bierregaard, eds. Chicago: University of Chicago Press.

Wilcove, D. S. 1994. Turning Conservation Goals into Tangible Results: The Case of the Spotted Owl and Old-Growth Forests. In *Large-Scale Ecology and Conservation Biology*, P. J. Edwards, R. M. May, and N. R. Webb, eds. Oxford: Blackwell Scientific.

Williams, C. B. 1964. *Patterns in the Balance of Nature*. London: Academic Press.

Williams, M. 1989. *Americans and Their Forests: A Historical Geography*. Cambridge: Cambridge University Press.

Williams, P. 1998. Key Sites for Conservation: Area-Selection Methods for Biodiversity. In *Conservation in a Changing World*, G. M. Mace, A. Balmford, and J. R. Ginsberg, eds. Cambridge: Cambridge University Press.

Wilson, E. O. 1992. *The Diversity of Life*. London: Penguin.

——— 1997. Introduction. In *Biodiversity II. Understanding and Protecting Our Biological Resources*, M. L. Reaka-Kudla, D. E. Wilson, and E. O. Wilson, eds. Washington, DC: Joseph Henry Press.

——— 2003. The Encyclopedia of Life. *Trends in Ecology and Evolution* 18:77–80.

Wilson, E. O. and F. M. Peter, eds. 1988. *Biodiversity*. Washington, DC: National Academy Press.

Part II

Understanding Biodiversity

3

Plato on Diversity and Stability in Nature

JUHANI PIETARINEN

The view that Plato was primarily interested in the nonempirical world of the
Forms and disregarded the "sensible" world of natural phenomena is one-
sided and misleading, if not entirely wrong. He was very much concerned
with what we call empirical nature, which he believed to be in a state of
constant change. "The things of which we naturally say that they 'are', are in
process of coming to be, as the result of movement and change and blending
with one another; we are wrong when we say they 'are', since nothing ever
is, but everything is coming to be" (*Theaetetus* 152e). The notion of change
involves a difficult conceptual problem: to say that something is changing
seems to imply that it both remains the same and becomes something else; a
growing tree is at every moment transforming into something else while all
the while remaining the same tree.

   Plato returns to the problem of change over and over again in his dialogues,
and in the *Timaeus* he offers an especially interesting explanation for changes
in the "sensible" world of nature. The explanation is rather complicated and
difficult to understand in detail. However, its general dynamical tenet has an
interesting affinity with our modern biological view of nature. According to
Plato, the essence of nature consists of some kind of universal power that
produces maximal diversity and abundance in the natural world and regulates
changes in accordance with uniform laws. The laws bring stability in the
constant flux of changing phenomena, and certain passages in the *Timaeus* can
be read as an explanation for the relationship between diversity and stability. I
make some comparisons of Plato's view with certain current views of stability
and its relation to diversity in ecological biology.

An earlier version of this chapter appeared under the title "Plato and Biodiversity" in K. Boudouris
and K. Kalimtzis, eds., *Philosophy and Ecology I*. Athens: Ionia Publications, 1998.

85

Another common view is that Plato disvalued nature in the sense of regarding animals and plants as inferior to humans and for the most part worthless for human purposes. This, too, is one-sided and in an important sense wrong. In the *Timaeus*, Plato develops a powerful argument to prove that the most valuable thing in the universe is maximal diversity, which implies, I will argue, that every kind of being has independent value because it adds something to this diversity. Human beings form one part of the total system of living beings, and because the living beings form a unity, a kinship of humans with animals and to some extent also with plants, must prevail. Most importantly, all living beings have rationality, at least to some extent, so that not even our rational powers offer grounds for separating us radically from the other parts of nature. Plato's view is of course that more of such powers are given to humans than to animals and plants, so in this sense we can be considered superior to the rest of nature. However, what follows from this superiority is not a right to suppress other living beings and treat them despotically but our responsibility for the well-being of nature as a whole. According to my interpretation, by this argument in the *Timaeus* Plato gives an interesting answer to the question of why we should protect diversity in the biosphere.

3.1 DIVERSITY AND STABILITY IN NATURE

I mainly refer to species count as a measure of diversity; it offers the simplest way of estimating diversity and corresponds best to what Plato meant by the plenitude of nature. The measure may be refined by taking the relative frequency of individual members of species into account, and this seems relevant for some questions arising from the *Timaeus*.

Stability, generally speaking, refers to the ability of systems to withstand changes. For instance, a community's ability to resist invasion can be taken as an indication of its stability. In ecological theory, definitions of stability in terms of the idea of equilibrium have been common: a stable system returns easily to its equilibrium after small perturbations, and system stability increases as time required to return to equilibrium decreases. The main problem with such definitions is the difficulty of specifying the notion of equilibrium in ecological systems. Constant competition for food as well as genetic variability cause continual changes in populations, and ecologists have adopted measures of stability which rely on variability in population or community densities (see McCann 2000). In general, stability increases as density moves further from extremely low or high densities, that is, when variation in density decreases. Recent experimental results indicate that diversity measured as species richness within an ecosystem tends to be positively correlated with stability measured as decreased variability in community density.[1] This, however, should not be taken to mean that diversity is the driver of this relationship, but that ecosystem stability depends on the ability of the species in the communities to respond differentially to perturbations. By regulating their interaction processes like food-web structures, rich ecosystems are able to protect themselves against internal and external destabilizing factors, and decreasing biodiversity tends to increase the overall mean interaction strength and thereby the probability that systems undergo destabilizing dynamics and collapse (McCann 2000, 233).

Whatever the basic mechanism here might be, the interesting general feature of stable ecosystems is that they have a kind of self-regulating or autogenic power to preserve their functioning by responding differentially to perturbations, and that such dynamic stability correlates positively with diversity.[2]

This, in effect, was also Plato's view.

3.2   PLATO'S EXPLANATION FOR CHANGES IN NATURE

The sensible world is a world of motion and change for Plato. What are these motions? How should they be explained? These questions motivated Plato to develop his famous doctrine of the Forms.

The realm of Platonic Forms might be said to represent an ideally ordered world, one in which everything fulfills the requirement of perfect harmony. Whatever things exist in such a world, they must consist of parts that are harmoniously related to each other, and also the relations between them must exhibit perfect harmony. Such things are of course entirely immaterial, purely objects of reason. Because the perfect order cannot change in any way, the realm of Forms is a world of eternal *being*, something that "always is and has no becoming" (*Timaeus* 27d). The supreme "power" producing harmonious order in this realm is called by Plato the Form of Good, probably because it is what makes everything, including all other Forms, perfectly good and beautiful.

If Plato had taken the realm of Forms to be the only reality, he would not have been able to explain motion and change. From where does motion come into our world? To answer this question, Plato postulated something entirely opposite to the world of the Forms, a primeval state of chaos consisting of entirely random motion or change (*Timaeus* 30a, 53b). It might be best to think of this realm, which Plato called *khora*, as referring to a principle or power of perfect disorder, as something forming the exact opposite to the Form of Good. Plato rejected as unsatisfactory the theory that the visible world would basically consist of a few material elements, for he realized that by simply postulating such elements, we cannot explain motions and changes such as fire becoming air and air becoming earth. He thought that something more basic than the elements was needed: a power causing disorder and thereby keeping things in constant motion. *Khora* represents such power; by it things *resist* all kinds of constant order and stability. Plato says that we cannot comprehend this kind of power as such because it has no independent existence but must "come to be in something else, somehow clinging to being, or else be nothing at all" (52c). *Khora* does not exist in itself but only in those things where its effects become apparent, that is, in moving and changing sensible things. As Plato says, "it takes on a variety of visible aspects" (52e). The power resisting order is a real constituent of the empirical world, it is inherently present in all sensible phenomena; as Plato says, sensible objects "participate" in it (50c). All motion and change are due to this power.

By means of the two basic principles, the ideally harmonious order and the perfect disorder, Plato explains the genesis and nature of our empirical world. Representing the power of the Form of the Good, Plato's creator (*Demiourgos*) tends to bring order into the intrinsically chaotic "proto-material," "because he believed that order was in every way better than disorder" (30a). However, the ordering power is resisted by the power of *khora* so that the result is always a compromise: the material bodies are constantly moving and changing but there is more or less regularity or order in their motions. The objects of the sensible world are always in the state of "becoming," that is, of moving from one state to another, yet this becoming is not entirely random but governed by regularities that reflect the ideal of harmonious order. The amount of order may vary considerably depending on how much things are influenced by the power of *khora* compared to the power of the Form of the Good. For instance, the motions of heavenly bodies are extremely regular, whereas the motions of material elements like fire and air are very unpredictable because they are heavily influenced by *khora*.

Plato describes in the *Timaeus* the role of the basic powers in the explanation of natural phenomena when he speaks of two different kinds of causes. "Rational" or "divine" causes refer to the power that brings rational order and stability into the universe and in this way produces "what is excellent and good" in natural phenomena (46e). The other kinds of causes, which Plato called "auxiliary" or "necessary," produce "only accidental and disorderly effects every time," that is, variability occurring in changing phenomena. A proper explanation of natural processes should pay attention to both of these kinds of causes.

To account for how the rational causes bring order into the empirical world, Plato introduces the notion of soul. To understand his idea of the soul, let us try to think of something that moves but is as close to the idea of harmony as possible for a moving thing. Such a thing belongs to the world of becoming because it is moving, but its motions reflect the harmonious order as perfectly as possible. According to Plato's story about the creation of the soul in the *Timaeus*, the Demiurge mixed the indivisible Being consisting of the "same," and the divisible Becoming consisting of the "different," an intermediate form of being "to make a uniform mixture, forcing the 'different' that was hard to mix into conformity with the 'same'" (35a). Being this kind of mixture of the same and the different, the soul is a moving thing, but unlike material things it has nothing irregular in its motion because it follows only the harmonious order of the Forms. But harmonious order is for Plato rational order, and when the Demiurge realized that nothing apart from the soul can be rational in the universe, he "put reason in soul and soul in body" (30b).

The important thing in this story is that the soul, following only rational laws, is capable of regulating its own motions. The reason is for Plato a perfectly self-regulating faculty. In the *Phaedrus* Plato argues for the immortality of soul by appealing to its capability of self-motion (245c–d), and he concludes that this kind of self-mover must be a source of all regular motion. But since we have found that a self-mover is immortal, we should have no qualms about declaring that this is the very essence and principle of a soul, for every bodily object that is moved from outside has no soul, while a body whose motion comes from within, from itself, does have a soul, that being the nature of a soul (*Phaedrus* 246e).

When the Demiurge put this kind of soul in a body, a special kind of being was created: a living being. For Plato, living beings are bodily beings possessing the capacity to move themselves by virtue of the soul's power (*Phaedrus* 247c). The natural world is a diversity of such living beings. Plato does not consider it as a static collection of various kinds of species but rather as a dynamic diversity where species and individual bodies evolve from existing ones, move and occupy different regions, appear and disappear leaving space for new ones. His view is in interesting respects reminiscent of our conception of nature as a network of ecosystems.

3.3 PLATO AND THE DIVERSITY OF THE NATURAL WORLD

Why is there diversity in the sensible world? Plato's answer to this question is important not only for understanding how he explained natural phenomena but also for obtaining a proper view of what he thinks of the value of nature.

Plato's starting point of the argument in the *Timaeus* for diversity is that the Demiurge was perfectly good, and we cannot permit "that one who is supremely good should do anything but what is best" (30a). Because order is better than disorder, the nature as a whole must have as much order as possible, and this implies that it should have a rational soul in its material body. Thus, Plato infers that the sensible world must be a living being (30c). But to be the best possible living being, it must become as closely as possible the ideal model of a living being. What is this model? It must be the Form of a living being, that is, the most perfect living being in the sense that it is not possible to conceive of any kind of living being that this Form is not able to represent ideally; Plato says that it comprehends within itself all intelligible living beings (30c). With this model in mind, the Demiurge put soul into the material world and tried to make this resemble the model of a perfect living being as much as possible. Wanting nothing more than to design the best of all intelligible things, complete in every way, he made of the sensible world a single living being containing within itself all other living beings of like nature (30d–31a).

In this way the sensible world contains all possible kinds of living beings so that it is endowed with the maximum amount of diversity. Because they have souls, all living beings are capable of self-regulation, but this capacity varies from one being to another. The natural world consists therefore of beings with all possible degrees of "livingness" or rationality. All of them are necessary for the perfection of the world as a whole, and each different being adds something, however little, to the goodness of the world. Plato classifies living beings roughly into five principal sections: moving celestial bodies or "the heavenly race of gods" (the Sun, the Moon, the planets), men, birds, wild land animals, and animals living in water; this is both the time order of the Platonic evolution and the order of descending rationality (*Timaeus* 40a, 91d–92b). These five sections form the primary "species" in Plato's system.

The variety of species is not the only thing that contributes to the perfection of nature, for we can read from Plato's argument that he also considers differences between members of one and the same species of great importance. Not only individual human beings differ from each other with respect to their rational capacities; individual animals may also live a more or less rational life even when belonging to the same species. This follows from Plato's assumption that "all other living beings are parts, *both individually and by kinds*," of the perfect living being used by the Demiurge as his model of the sensible world (*Timaeus* 30c; emphasis added). Plato's idea seems to be that the extent to which the power of *khora* is able to resist the ordering power of the soul varies between different bodies of the same general kind; like a smith may forge more or less perfect exemplars of scythes, the quality of the individual living beings produced by Plato's Demiurge varies from one representative of a species to another.

This is an important point. For one thing, it allows Plato to overcome inter-species differences and compare individuals with respect to their rationality alone, irrespective of the species to which they belong. He remarks, for instance, that some human beings may be so simple-minded that they deserve the status of birds, and the least rational and ignorant ones belong to the level of the lowest water animals (91d–92b). On the other hand, the best individuals of animal species may reach the level of human rationality, at least the average level though not the highest one.[3] For Plato, the doctrine of interspecies transformation, according to which "the animals exchange their forms, one for the other, and in the process lose or gain intelligence and lack of it" (92c), was intelligible for Plato. I will return to this doctrine later.

Another consequence of the intraspecies diversity is that it makes possible for individual members to make their own contribution to the perfection of the sensible world. The Demiurge produces rational order in the world by means of the souls of living beings, and the creation of the orderly world continues all the time through individual souls, because they endeavor to advance the good purposes of their creator. This is not only true of men but also of other living beings, although we are to a greater extent responsible for our conduct than beings endowed with less rationality. However, all living beings have their place and function in the world: they contribute to the goodness of nature with the amount of self-regulating power they can exhibit, and the more individual beings are able to exercise this kind of power, the better world will be developed.

According to the picture outlined in the *Timaeus*, the empirical world is a plenitude of living beings representing all degrees of rationality or self-regulating power. Its rich diversity consists of inter- and intraspecies variation, and because of transformations from one species to another, there appears variation in the relative abundance of species. But does the idea that the world is a plenitude of living beings allow for any variation in the number of species from one time to another? Would Plato have thought extinctions of existing species or generations of new species possible?

One of the aims of Plato in the *Timaeus* is to "trace the history of the universe," first down to the emergence of humankind and then to the generation of other living beings. According to his story, new species have been developed over time. There is nothing in the principle of plenitude that would require all possible kinds of beings to exist at the same time; it implies only that each of them will exist *at some time*. However, there is no evidence that this is how Plato understood the plenitude of our world. He concludes the *Timaeus* by saying that "our one universe, indeed the only one of its kind, had come to be," and that "its grandness, goodness, beauty and perfection are unexcelled" (92c). Plato thought that the creation of new species had been completed, and that the world would remain in this state of perfection: although new species have been generated during the evolution of the natural world, it is not possible that any species could be generated after the completion of the creation. Nor could Plato leave any chance for extinctions, for it would mean a decrease in the perfection of the world and thus be against the good intentions of the Demiurge. We can conclude therefore that in Plato's view the number of species in our natural world will remain constant.

3.4   PLATO AND THE ECOSYSTEMS

According to Plato's argument, our world, being as rich in life as possible, is the best of all possible worlds. "Since the god wanted nothing more than to make the world like the most beautiful and perfect of the intelligible things, he made it a single visible living thing, which contains within itself all the living things whose nature it is to share its kind" (*Timaeus* 30d). The beauty and perfection of our world is not only based on the fact that it contains all kinds of living beings, but also on its *working* as much as possible like a perfect living being, that is, on the capability of self-regulated motion exhibited by it and its parts.

Plato describes the created universe as a whole as a giant ecosystem. It is a self-sufficient system resistant to all disturbances from outside:

> For since there wasn't anything else, there would be nothing to leave it or to come to it from anywhere. It supplied its own waste for its food. Anything that it did or was affected by it was designed to do or suffer by itself and within itself. For the builder thought that if it were self-sufficient, it would be a better thing than if it required other things. (33c)

Although all of its parts are in a constant flux, nature as a whole is a perfectly self-regulating system. We can identify in the *Timaeus* certain principles which are meant to account for the regularity and stability in the changing states of the natural world.

1. *Every empty space in nature tends to be fulfilled.* Plato explains the heterogeneity of the material world in terms of the pressure, caused by the circular motion of the universe, that constricts the elementary material particles "and allows no empty space to be left over" (58a). This kind of process forces the particles to move into new regions. The same is also true of bodies consisting of elementary particles, so that "the occurrence of heterogeneity is perpetually preserved, which sets the bodies in perpetual motion, both now and in the future without interruption" (58c). Occupying an empty space a body occupies a new region, and another body fulfills in turn the region left empty, so that any motion of a particular body causes a motion in the whole world. Because this kind of process is more or less irregular, we have to look primarily for "auxiliary" and not for rational causes to explain them.

2. *A thing or system in nature is the more stable, the more regular and self-determining it is.* The regularity of motions and the ability of things to determine their motions themselves depend on rational causes, that is, on the ability of living bodies to follow rational principles of motion. "The best of the motions is one that occurs within oneself and is caused by oneself, for it is the motion that bears the greatest kinship to understanding and to the motion of the universe" (89a). The more power to self-regulate a thing has, the less it is exposed to external causes and the more stability and permanence it possesses.

3. *A natural thing or system that has achieved a relatively stable state, tends to return to that state after disturbances.* Plato regards stable conditions good and disturbances bad. "An unnatural disturbance that comes upon us with great force and intensity is painful, while its equally intense departure, leading back to the natural state, is pleasant" (64d). By a disturbance he means that something is added or taken away from a complex system (like a living body), and a system can remain as "stable, sound and healthy" only when "that which arrives at or leaves a particular bodily part is the same as that part, consistent, uniform and in proper proportion with it" (82b).

Plato never explicitly formulated these principles, but they are plain enough in the *Timaeus*. The last two are used by Plato to account for various functions of the human body, but he could have widened their scope to ecosystems, for it is characteristic for Plato to consider complex systems as superbodies composed of several smaller bodies. For instance, he regards human bodies as systems consisting of smaller parts and these again of smaller ones, and the whole universe as a superbody that is composed of an infinite variety of material parts governed by the same causes of material necessity and rational regularities. Nature is for Plato a self-regulating "ecosystem," and as a whole it is completely stable in the sense of being free of external disturbances. It is entirely "autogenic," to use Bryan Norton's word. Similarly, the motions of the heavenly bodies remain stable because they are outside the influence of external causes. In contrast, all parts of what we call the biotic world are exposed to disturbances, for a limited system can never be entirely self-regulating.

We may plausibly think of Plato as granting that the principles stated previously apply to all natural systems. Any part of nature, however limited the ecosystem, is at least to some extent able to regulate its functions. According to principle 2, the stability of a system depends on this ability to self-regulate. Plato here comes quite close to the modern idea of dynamic stability that correlates, as recent findings seem to show, in a significant way with biodiversity. Principle 1 proposes that members of new species tend to fulfill empty spaces after perturbations. Without qualification, this is not an acceptable ecological "law," but it suggests that self-regulation and diversity are closely related: a limited system is the more capable of self-regulation the more variety it contains, that is, the closer it comes to the richness of nature as a whole. Plato's explanations may differ in many respects from present scientific views, but his main insight is still valid: biodiversity and ability to self-regulate are highly significant in the functioning of ecosystems.

According to principle 3, a system that has reached a relatively stable condition tends to return to this condition after disturbances. The idea of stability as a kind of equilibrium is still central in recent ecological discussions, in spite of the difficulties involved in the concept of equilibrium. Such notions as "constancy," meaning the ability of a system to resist all change when disturbed, and "resilience," understood as the ability of a disturbed system to return to its normal state after disturbances, are useless because ecosystems typically respond to disturbances by developing new points of stability in response to new disturbances.[4] Moreover, ecosystems undergo natural changes, such as adaptations of species to other species, and also genetic adaptations of species through evolution. Such systems are neither constant nor resilient, and, as discussed above, ecologists have developed measures of stability in terms of variability in population or community densities. As to Plato's position, it is clear that he would not have considered limited ecosystems as constant, that is, as resistant to changes. The heavenly bodies alone constitute a constant system, but otherwise the natural world is undergoing changes "without interruption," as he says. He could more plausibly hold the resilience view: he thinks that all systems have a "healthy" and "good" state, and in this sense a natural or normal equilibrium to which they tend after perturbations. However, irrespective of what the exact interpretation might be, the basic idea expressed by principle 3 is very important and useful in present-day ecology.

In sum, a natural system is for Plato the more self-regulating the more different kinds of species are involved in it, and with increasing diversity it approaches the perfect stability that is characteristic for nature as a whole. In this way biodiversity increases stability. Plato takes self-regulation as the most wonderful feature of nature as a whole: "This world of ours has received and teems with living things, mortal and immortal. A visible living thing containing visible ones, perceptible god, image of the intelligible Living Being, its grandness, goodness, beauty and perfection are unexcelled" (*Timaeus* 92c).

3.5   OUR RESPONSIBILITY FOR NATURE

By creating the soul, the Demiurge made the material world into something very important and unique: an active power that causes regularity and order there. The soul makes of material bodies living beings which, by their souls, participate in the rational harmony of the Forms. As mentioned previously, Plato takes this active power to be reason, and reason is for him the only power by which true knowledge can be obtained.

Plato believed that human reason, if properly exercised, is capable of understanding in the right way the basic laws of nature and their relation to the perfectly rational realm of the Forms. All understanding is based on the ability of the soul to move itself.

> The best of the motions is one that occurs within oneself and is caused by oneself. This is the motion that bears the greatest kinship to understanding and to the motion of the universe. Motion that is caused by the agency of something else is less good. Worst of all is the motion that moves, part by part, a passive body in a state of rest, and does so by means of other things. (*Timaeus* 89a)

As we have seen, Plato does not think that humans are the only beings in the natural world that are endowed with rational capacities. All living beings have it to some extent. According to Plato, it is the soul that makes birds pursue upward, toward the heavenly bodies, but they cannot rise high enough because the heavenly gods can only be reached by a soul completely free of the limitations of the senses, only by pure reason and understanding (91d–e). The souls of wild land animals are much more bounded by their material bodies, and fish and other water-inhabiting living beings have the least degree of rationality among animals (91e–92b). Interestingly enough, Plato also counts plants among living beings. Although "not entrusted with a natural ability to discern and reflect upon any of their own characteristics," they nevertheless "share in sensation, pleasant and painful, and desires" (77b). Plants are sentient but not self-conscious.

Plato made important moral conclusions from the view that living beings are capable of self-regulation. Because the rationality of the soul, reminiscent of the harmonious order of the Forms, represents the best that the natural world can exhibit, then every living being must have inherent value and therefore be morally significant at least as moral patients.[5] However, Plato also accepts the stronger conclusion that living beings, because of their rationality, are also responsible moral agents, although in proportion to their rational capacities. Plato seems to think seriously that all living beings should be held responsible for their behavior, for the way they live their relatively short lives. At the end of the *Timaeus*, Plato explains that different kinds of animals are transformations of men who have not developed their rational abilities or have used their reason in the wrong way. Birds have descended from simple-minded but benevolent men, wild land animals from men who have paid no interest in heavenly things but followed only their bodily desires, and the most ignorant men have been transformed into water animals, for "their souls were tainted with transgressions of every sort" so that they no longer deserved to breathe pure air (91e–92b). But Plato goes further and says that such transformations are not only the fate of irrationally living men but "all the animals exchange their forms, one for the other, depending on the loss or gain of understanding or folly" (92c). This kind of generation of species from others is an ongoing process; it occurs, Plato says, "both then and now," from the beginning of the creation onward.

Note that the transformations depend on the loss or gain of understanding. The traffic is two-way: "Each soul chooses the life it wants," Plato says in the *Phaedrus*, so that "a human soul can enter a wild animal, and a soul that was once a human can move from an animal to a human being again" (249b–c). This means that members of any species may rise to a higher species or fall to a lower one depending on how they choose to use their rational abilities. Thus, the requirement of moral development concerns all living beings and not only humans, and the individuals ignoring it have to try again as members of a lower species. The idea of animals as moral agents choosing their way of life sounds strange, and Plato's point may be taken just as a warning for people. However, we have to remember that it is a logical consequence of Plato's starting point, and therefore it was natural for him to think that all living beings, who have at least some rationality, are responsible for their own fate, though vanishingly little in most cases.[6]

This leads to an interesting dilemma that Plato seems to have ignored. If all individuals are responsible for the use of their rational capacities, and if individual behavior in this respect occasions interspecies transformations, then the diversity in the sensible nature becomes to some extent dependent on how individuals conduct their lives, in other words, on what kind of moral choices they make. The variety of the world is not fixed to the last feature, but some room has been left for the inhabitants to shape it in accordance with their preferences. From the viewpoint of moral responsibility, the individual members of all species are obliged to perfect themselves by exercising their rational capacities as fully as possible. But this, in turn, means that there should occur only upward transformations between species until the lower kinds of living beings disappear and at the end only the most rational human beings (the Platonic philosophers) remain. The requirement of moral perfection seems to imply decreasing diversity in the world, both in the interspecies and intraspecies sense.

Human beings are here in the decisive position, because we are capable of living in the most responsible and self-determining way. If Plato was serious with his view of the down- and upward transformations, he should have realized that the requirement of rational perfection can be fulfilled only at the cost of natural diversity. John Passmore (1974, 33) notes that the Greek view in general implies that our task of perfecting nature amounts to "humanizing" it, making it "more useful for men's purposes, more intelligible to their reason, more beautiful to their eyes." But in view of Plato's teaching of animals as descendant humans, this can happen only by educating people to fully use their rational potential, which, if successful, would lead to a radical impoverishment of natural diversity. Is this what Plato should expect a humanized nature to be like?

Of course not. To think so would be strongly against the story told in the *Timaeus* of the creation of the sensible realm. I have argued that according to Plato's considered view, all kinds of living beings have an important function in the natural world, for they bring into it at least some regulating power and contribute thereby to the goodness and perfection of the world. The Demiurge created beings of all degrees of perfection, according to the model of the Living Being that "comprehends within itself all intelligible living things, just as our world is made up of us and all the other visible creatures" (*Timaeus* 30c–d). The clear implication of this is that the grandness, goodness, and beauty of this world consist of its richness, that is, of the great diversity of species and individuals. Once this kind of world has been evolved, it will remain in its state of maximal richness.

If this is Plato's considered view, what should we say about the transformation doctrine that seems to conflict with it? One possible answer is that the doctrine was not meant as a serious biological explanation but should be understood merely as a moral teaching. But what grounds do we have for thinking so? In the *Timaeus* Plato offers his most ambitious explanations for natural phenomena, and nothing indicates that he did not mean the transformation doctrine to be taken seriously. It seems more plausible that Plato simply ignored the dilemma arising from it. There is a natural explanation why the dilemma did not occur to him: Plato was extremely pessimistic about successful moral education of the majority of people. "I suppose that everyone would agree that only a few natures possess all the qualities . . . to becoming a complete philosopher and that seldom occur naturally among human beings," he remarks in the *Republic*, and adds that even this small, naturally talented elite is liable to inclinations like wealth and power (491a–c). Thus, if only a few persons will meet the requirement of moral perfection, and the living beings of nonhuman nature are less capable of perfecting themselves than most humans, the requirement cannot be a real risk for natural diversity. Plato could by no means think that human perfection would in any way be problematic with respect to diversity; rather he saw the risk in our lack of rationality, for the following reason.

According to Plato, "it is everywhere the responsibility of the animate to look after all that lacks a soul" (*Phaedrus* 246b). This I think should be read primarily as a moral norm. We humans have received a self-determining soul and must therefore look after things endowed with less rationality. This means, as Plato often emphasizes, that we are responsible for our own bodies, for their health and well-being, but it also means that we are responsible for those with less rationality. We are the rulers of animals and plants, and Plato compares good rulers with shepherds caring for their herd: good shepherds and cowherds seek the good of their sheep and cattle, and not their master's or their own good (*Republic* 343b). Whatever it may mean, in a more detailed analysis, to look after the good of nature and not just our own good, it is obvious that we should take care of the natural diversity. We have seen that for Plato, at least according to *Timaeus*, the best thing in the sensible realm is the great diversity of living beings, and hence the worst thing to do would be to impoverish the diversity by reducing the frequency and life space of the members of the other species.

When Plato speaks of our own bodily and mental well-being, he says that "there is but one way to care for anything, and that is to provide for it the nourishment and the motions that are proper to it" (90c). When this is applied to nature in general, we care best for the well-being of other living beings by providing nourishment and motions appropriate for them. To accomplish this, we have to leave enough natural environment for species to flourish in all the diversity meant by the perfectly good great Demiurge.

NOTES

1. This is McCann's summarizing statement based on about two dozen research reports; see McCann 2000, 229–31.
2. Bryan Norton (1987, 80–81) proposes that a dynamic stability he calls "autogenic" might correlate with diversity.
3. Gabriela Roxana Carone points out correctly that it is not humans as a species that Plato regards as superior to animals, but the rational soul of the best of humans (the philosophers); see pages 123–4 of her important paper "Plato and the Environment" (Roxana Carone 1998).
4. For a useful discussion of these notions, see Norton 1987, 31–2.
5. The distinction between moral agents and patients is commonly used in current ethical discussion. Moral agents have the ability to make self-conscious moral decisions and hold themselves responsible for their actions. Moral patients are beings toward which moral agents have duties and responsibilities, but they need not be capable of making moral decisions. Moral patients may cause benefit or harm but they cannot be said to do right or wrong things. The important thing is that moral patients can be treated well or poorly by agents who are always responsible for the benefit or harm they cause (Regan 1988, chap. 5, section 5.2).
6. Roxana Carone (1998, 123) claims that for Plato ordinary humans (apart form philosophers) and nonhuman animals are not different in any relevant way but have equal possibilities of choice and "fall and rise." This is so because the only thing that matters, in Plato's view, is the rationality of living beings, not to what species they belong. In other words, Plato takes the intraspecies variability of rationality of humans to be as high as the variability among the different species, and in this sense humans and animals can be said to have equal possibilities of choice. On the other hand, as Roxana Carone also points out, Plato emphasizes that only humans can achieve a full exercise of reason and become a moral agent in the proper sense.

BIBLIOGRAPHY

Carone, G. R. 1998. Plato and the Environment. *Environmental Ethics* 20:115–32.
McCann, K. S. 2000. The Diversity-Stability Debate. *Nature* 405:228–33.
Norton, B. G. 1987. *Why Preserve Natural Variety?* Princeton, N.J.: Princeton University Press.

Passmore, J. 1974. *Man's Responsibility for Nature*. London: Duckworth.

Plato. *Phaedrus*. Trans. A. Nehemas and P. Woodruff. In *Plato. Complete Works*, J. M. Cooper, ed. Indianapolis, Ind.: Hackett, 1997.

Plato. *Theatetus*. Trans. M. J. Levett, rev. M. Durnyeat. In *Plato. Complete Works*, J. M. Cooper, ed. Indianapolis, Ind.: Hackett, 1997.

Plato. *Timaeus*. Trans. D. J. Zeyl. In *Plato. Complete Works*, J. M. Cooper, ed. Indianapolis, Ind.: Hackett, 1997.

Regan, T. 1988. *The Case for Animal Rights*. London: Routledge.

4

Biodiversity, Darwin, and the Fossil Record

KIM CUDDINGTON AND MICHAEL RUSE

Some ideas about nature owe more to our historical beliefs than to current theories or data. Because nature has been seen as an expression of the mind of the divine, it is particularly prone to be interpreted in light of prior beliefs. As it was once a heresy to suggest that the earth circles the sun, so in many quarters it is still taboo to suggest that the natural state of our global ecosystem is not one of stability. We will argue that this belief is something rooted more in our religious past than in our scientific present. Adopting a common usage of the term "biodiversity" to mean the total number of species in an area, we address Darwin's ideas on biodiversity and the claims of modern interpreters of the fossil record. We claim that these ideas are influenced by older prescientific concepts regarding the bounty and balance of nature.

Darwin firmly rejected the idea that the existence of particular identities of species was fixed. He claimed that species went extinct and others came into being. Yet, Darwin and modern authors have interpreted the fossil record as supporting the hypothesis that the total number of species will reach an equilibrium value. We shall demonstrate that the basis of this claim, and a similar claim presented by modern authors, is weak. This emphasis on equilibrium interpretations may be due to an idea prevalent in the seventeenth century regarding the state of nature. That is, in early modern Western history, nature was interpreted as reflecting the influence of a divine creator. In the earliest interpretations, it was believed that the number of species was fixed and immutable. However, when Darwin demonstrated this was not the case, the predictability of nature was then expressed in the claim that although individual species may go extinct, the number of species would tend toward some constant value. We claim that both Darwin's interpretation, and that of modern authors like Jack Sepkoski, are influenced by these cultural ideas, and that other interpretations of the fossil record are clearly possible.

To demonstrate that scientific hypotheses have been accepted for reasons other than those that are justifiable in a scientific context (e.g., empirical demonstration, logical coherence, simplicity, fruitfulness) is difficult, especially when such claims have been made more than a hundred years ago. We rely on several lines of evidence to support our claim. First, both visually and statistically there is little evidence that an equilibrium in species number has been reached. Second, the modeling framework used by both Darwin and modern authors to support the claim that species number is at equilibrium does not necessarily make that prediction. Finally, several supporting arguments are not logically consistent with prior claims. Given these three arguments, we suggest then that both Darwin's claim and some modern authors' claims may owe more to cultural beliefs about natural systems than to fossil data or logical arguments.

4.1 DARWIN'S EQUILIBRIUM MODEL OF SPECIES DIVERSITY

Prior to Darwin, it was believed that the face of nature reflected the mind of God. A given number of species had been created, and were providentially endowed with reproductive qualities that compensated for natural disasters, predation, and disease. Species extinctions were not possible, since nature reflected the perfection and benevolence of the deity (Egerton 1973). (Editors' note: This presumption originates from Plato; see Pietarinen's chapter in this volume.) With the explorations of the Comte de Buffon and of Georges Cuvier, scientists began to realize that species extinctions had occurred in the past. It was Charles Darwin, however, building on the thinking of the geologist Charles Lyell, who was the first to try to put this idea into an overall conceptual framework, backed by a substantial body of convincing evidence. Darwin made a clear claim that species went extinct, and further that new species came into being. But at what rate? In the arguments presented in the *Origin* on the role of competition and natural selection, there appears at first glance no reason to suppose there should be a particular number of species. In the first edition of *Origin*, however, Darwin claims that there is a balance between the rates of speciation and extinction. He states, "it inevitably follows, that as a new species in the course of time are formed through natural selection, others will become rarer and rarer and finally extinct" (Darwin 1859, 110). This balance between the two rates produces, at least eventually, a dynamic equilibrium in the number of species:

> In flourishing groups, the number of new specific forms which have been pro-
> duced within a given time has at some periods probably been greater than the

102

number of old specific forms which have not been exterminated, but we know that species have not gone indefinitely increasing at least in the later geological epochs, so that looking to later times, we may believe that the production of new forms has caused the extinction of about the same number of old forms. (Darwin 1872, 326)

When a new species comes into existence, an old species goes extinct. Therefore, we expect there to be about the same number of species at all times.

Unsurprisingly, Darwin claims that the underlying mechanism that produces this balance between extinction and origination is natural selection due to competition. Generally, he suggests that this balance is achieved by newly derived forms exterminating their closest relatives. He says, "each new variety or species during the progress of its formation, will generally press hardest on its nearest kindred and tend to exterminate them" (1859, 110), and "hence the improved and modified descendent of a species will generally cause the extermination of the parent species" (1872, 326).

Rather than envisioning natural selection supporting the radiation and accumulation of many new species, Darwin views its role as that of maintaining an equilibrium. He says:

> The theory of natural selection is grounded on the belief that each new variety and ultimately each new species, is produced and maintained by having some advantage over those which it comes into competition; and consequently extinction of the less-favoured forms almost inevitably follows. (1872, 326)

In other sections, Darwin expands his explanation of the mechanisms determining species number. He further suggests that there are physical limits to the number of species in a given area:

> What then checks an indefinite increase in the number of species? The amount of life (I do not mean the number of specific forms) supported on an area must have a limit, depending so largely as it does on physical conditions; therefore, if an area be inhabited by very many species, each or nearly each species will be represented by a few individuals; and such species will be liable to extermination from accidental fluctuations in the nature of the seasons or in the number of their enemies. The process of extermination in such cases would be rapid, whereas the production of new species must always be slow. (1872, 128)

Here, two factors are claimed to control species number. In the first part of the passage, Darwin suggests that a physical area can only support a finite amount of biomass. This is certainly true. The current energy resources and the amount of physical space available for species are limited. Competition for these limited resources would drive some species to extinction. Darwin claims that when species number becomes larger than can be reasonably supported by these physical limitations, the rate of extinction will further exceed the rate of species origination "which will always be slow." Thus, species number will be brought back to equilibrium.

Darwin combines a second mechanism with the first in this text. In the second half of the passage, he suggests that since the amount of biomass supported by a given area is finite, a large number of species would necessarily entail a small number of individuals in each species. If only a few individuals represent a given species, it would most certainly go extinct due to unpredictable fluctuations in the weather, or the effects of competition and predation (i.e., demographic stochasticity).

Barry Gale (1972) interprets Darwin as breaking from older ideas about the balance of nature. He suggests that earlier naturalists, like Archdeacon William Paley, subscribed to a cosmology where nature was "harmonious, ordered, economical and plentitudinous." Darwin, on the other hand, is described as promoting a view of nature as "nondesigned, nonbeneficent, and imbalanced" (Gale 1972). However, Darwin's claim that an equilibrium number of species has been reached seems also to rest on ideas of plenitude and balance. The idea of plenitude is especially important given the key role assigned to competition. This is a pre-Darwinian idea that life fills all possible niches. The natural world is thought of as closely packed with species. In an early private notebook, just at the moment of discovery of natural selection (at the end of September 1838), Darwin drew an analogy between the number of species on the earth and a surface covered with "ten thousand sharp wedges." A new wedge could only be driven in by expelling another; origination of new species could only occur by displacing a pre-existing one. Darwin has replaced the idea that there are a fixed number of species with the claim that a dynamic equilibrium is maintained with species coming and going, but with constant total species diversity. Although the constancy of nature has been overturned, the bounty and predictability of nature have been retained.

It is not clear whether the notion of the fullness of nature drove Darwin to emphasize the role of competition, or whether his perception of the importance of competition lead to the assumption that species were tightly packed. This emphasis leads Darwin to underplay the role of physical factors in natural selection. However, the assumption of bounty and the importance of competition for controlling the number of species was clearly suspect at least for particular geographical locations. Darwin notes: "We do not know that even the most prolific area is fully stocked with specific forms: at the Cape of Good Hope and in Australia, which support such an astonishing number of species, many European plants have become naturalised" (1872, 97). Although, to be fair, Darwin does cite other instances of the displacement of native species by invaders which causes a turnover of species rather than an increase in number.

The strength of Darwin's argument rests on the claims that: (1) new species are represented by only a few individuals, and (2) extinction rates will increase with species density. It is certainly true that for very small population size, extinction is likely. However, Darwin's arguments on this topic will only hold in that case where species population size is being limited by a competition for resources. If one admits that biomass has not reached its maximum conceivable limit and new energy and space resources (not to mention niche spaces) will be created with the formation of new species, then there is no compelling reason to assume that a large number of species will be represented by a very few individuals of each kind.

So, for Darwin, species number will not increase continually, nor will it fluctuate substantially. Instead, the rates of extinction and origination are balanced to produce an equilibrium number of species. This number is a maximum that the area can support. But this is also apparently an issue on which he has not thought deeply. He states elsewhere that "the extinction of a whole group of species is generally a slower process than their production" (1872, 324) which if this were always the case, would imply that species number should continually increase. Further, the most speciose region he is aware of, the Cape of Good Hope, is not fully packed with species.

4.2   A MORE FORMAL INTERPRETATION OF DARWIN'S EQUILIBRIUM MODEL

Darwin has given us a general model of species number based on both competition and environmental constraints. The more species there are, in an environment with a fixed amount of energy and space, the more competition there is, and the more likely it is that species will go extinct. We can construe these statements as an equilibrium model of species number. Indeed, it is not apparent that there is any other possible interpretation. Darwin seems committed to the claim that species origination will always be slow; therefore, we can describe his model as a diversity-dependent extinction model of species biodiversity. That is, speciation rates are a slow constant, while extinction rates will vary with the physical resources available and the number

Model of Species Diversity

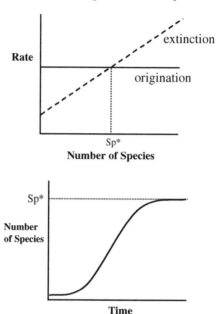

Figure 4.1. A mathematical interpretation of Darwin's diversity-dependent extinction model. Top panel: The constant rate of speciation and increasing rate of extinction with species number intersect at an equilibrium number of species (Sp*). Bottom panel: In time, the number of species will increase to this equilibrium.

of species depleting these resources. As a region becomes more speciose, we would then expect an increase in per-species extinction rates.

If speciation is always constant, and extinction is an increasing function of species biodiversity, the two rates may intersect to produce an equilibrium species number (Fig. 4.1). If one describes speciation and extinction as continuous-time events, then there are only a few possible model predictions. It seems likely that Darwin, if he envisioned his equilibrium model in mathematical terms, would have restricted his reflections to a linear increase in extinction rate with increasing biodiversity. An intersection between the two rates, less formally described by Darwin, produces an equilibrium number of species. When disturbed from this attractive state, rates of extinction or speciation will act to restore the balance.

Darwin does comment elsewhere on greater speciation rates associated with larger numbers of species, but does not seem to interpret this idea as an increasing per-species speciation rate with increasing biodiversity. Clearly, an increasing rate of speciation with increased biodiversity admits the possibility that there will be no intersection between the two rates, and thus no equilibrium number of species.

In the sixth edition of the *Origin*, Darwin responds to criticisms of this simple equilibrium model of species number. He notes that if one believes that new species create new opportunities for still later species, one might also assume that as the number of species increases, so does the ability of the system to support a larger number of species. He says:

> I fully admit that the mutual relations of organic beings are more important, and as the number of species in any country goes on increasing, the organic conditions must become more and more complex. Consequently, there seems at first sight no limit to the amount of profitable diversification of structure, and therefore no limit to the number of species that might be produced. (1872, 128)

Indeed, a simple modification, of increasing speciation with increasing diversity, may produce speciation and extinction rates that do not intersect, and an ever-increasing number of species.

Darwin's arguments about the limitation of physical space and energy are presented to counter the claim that speciation will increase with diversity, and seem reasonably sound at first glance. While species number may increase for a time, the hard boundaries of environmental conditions will eventually prevent any further increase in species number. We might describe this as a nonlinear relationship between speciation and biodiversity: an initial increase in per-species speciation rates at low biodiversity, followed by a decrease as we approached the limits of the environment. In this case we would, under many conditions, obtain an equilibrium number of species, although not necessarily so.

While there is a finite amount of energy or physical space over a given area at a particular point in time, it does not necessarily follow that these quantities are fixed. The earth is an open energy system in that sunlight is continually captured by plants and released back to the atmosphere as heat. We need only to ponder forest landscapes to remind ourselves that the amount of available energy on the earth has probably increased through time. Darwin himself comments on the increased productivity of a plot of land planted with many species (1872, 114), so he must have subscribed at least in part to the idea that as the number of species increased, more energy was available in the system.

Similarly, from an organismal perspective, the amount of area available for occupation will also vary through time. Unless living biomass is packed uniformly without spaces in a cylinder rising from the nonliving substrate, we will not have truly reached a maximum of physical space available for occupation.

Further, most species, by their presence, produce additional physical locations. A colonizing bush creates new physical space that can be occupied by insects and birds. Earthworms create tunnels that foraging roots will occupy and so on. An increasing amount of physical space or energy in the system would continually shift the balance between speciation and extinction, yielding no discernible equilibrium state until some hard physical limitation had been reached.

Clearly, the argument that Darwin presents is neither very detailed nor well defended. Competition is claimed to be the major determinant of species number. This claim, however, rests on the ideas that nature is tightly packed with species, and further, that new species do not provide a significant opportunity for further diversification. Michael Ghiselin (1995) points to Darwin's overemphasis on competitive relations as a symptom of his Victorian milieu; however, we suggest that, with respect to his ideas about species number, it results from both a desire to protect his larger theory from criticism, and contemporary beliefs about the orderliness of nature. It seems likely that in protesting the implication that natural selection should produce a continually increasing number of species, Darwin is in fact defending himself from the claim that his theory makes a nonsensical prediction. An ever-increasing number of species is an impossibility. On the other hand, there was no reason a priori for Darwin to suggest that this limit on the number of species had already been reached, nor to insist that species steadily accumulated toward this number. That is, while Darwin rejected the idea that species were eternal, he accepted the claim that the underlying character of the natural world was constant, and in its present state, bountiful.

4.3 DARWIN'S APPEAL TO THE FOSSIL RECORD

Darwin points to the fossil record to support his claim that in modern times there exist a more or less constant number of species. He says, "geology shows us, that from an early part of the tertiary period the number of species of shells, and that from the middle part of this same period the number of mammals, has not greatly or at all increased" (1872, 128). In particular, he suggests that the number of species of mammals in Europe has reached a maximum. That is, Darwin is not only claiming that the number of species in a given area should reach an equilibrium, but that it has already done so in some areas. In another text, Darwin suggests that while early parts of the fossil record do not support this claim that the number of species is maintained at equilibrium, that latter part of the record does. Darwin paints a picture of nature as essentially stable and predictable in his time, whatever it has been like in the past. Unfortunately, the evidence he relies on is not presented, while his arguments are not conclusive.

It may be that Darwin himself had counted the number of new species in later fossil records and compared it to the number of known species. However, given the number of new species he identified on scientific expeditions and the poor quality of the fossil record, it seems unlikely that anyone could conclude this was definitive evidence that the number of species in a particular group had reached an asymptote. Elsewhere, Darwin comments extensively on the problems of the incompleteness of the fossil records.

It seems more likely Darwin thought that a continuously increasing number of species was an absurdity, and he needed to convince others that this unreasonable outcome was not a prediction of his theory. The mechanism for natural selection is the same mechanism that controls species number: competition. Therefore, a nonsensical prediction regarding species number resulting from this mechanism would cast doubt on his theory as a whole. One can understand why a continually increasing number of species seems an impossibility, at least in infinite time. However, there is really no reason to suggest that an equilibrium number of species has already been reached; Darwin was also committed to the claim that speciation was a very slow process, requiring vast quantities of time. One also should note that there are other possible relationships not considered by Darwin: the number of species may fluctuate randomly in finite time, cycle, or increase and then decrease toward a lower number than that currently existing.

Given the paucity of evidence, and some contradictions in thought, it may be that Darwin thought there were a constant number of species in modern times, and constructed reasons to support that view. That is, Darwin was influenced by an earlier view that nature established some kind of equilibrium state. We speculate that Darwin was influenced by cultural ideas about the constancy of nature, but instead of fixed species, he postulated a dynamic equilibrium number of species.

4.4 MODERN INTERPRETATIONS OF BIODIVERSITY AND THE FOSSIL RECORD

We find Darwin's position restated in work by modern authors such as Michael Rosenzweig (1995) and Jack Sepkoski (1978, 1979, 1984). Although the merits of equilibrial and nonequilibrial models were actively debated in the 1980s (Carr and Kitchell 1980; Sepkoski 1984; Miller and Sepkoski 1988 vs. Walker and Valentine 1984; Cracraft 1985; Hoffman 1985; Benton 1987, 1997), even opponents agree that the mainstream position is that the global number of species has reached an equilibrium number. This interpretation is presented in several textbooks (Benton 1999). Although the exploitation of a new niche space or resources may have increased the equilibrium number of species a few times (see Rosenzweig 1995), modern species are claimed to be more or less at equilibrium diversity. Sepkoski, in particular, gives an account of species number as a result of balanced speciation and extinction rates that are dependent on the total number of species.

4.5 SEPKOSKI'S LOGISTIC MODELS

Sepkoski (1978) adapted the logistic model of population growth to give a qualitative prediction of the species accumulation curve for marine bivalves. Like some earlier authors, he suggested that the rates of species origination and extinction are dependent on species diversity, and intersect to produce an equilibrium species number. This model gives the same prediction as Darwin's verbal model. The equilibrium is a globally attractive state. When there are fewer species than the equilibrium number, more accumulate, when there are more species than the equilibrium, species are lost.

Adaptation of the logistic model for this purpose requires that one assumes that a species (or other taxonomic group) operates more or less as a single individual in a population. Following this, the assumptions of this use of the logistic model are then similar to those of the logistic model of population growth. Changes to species number occur more or less instantaneously; all species are competing for limited resources, which are fixed in number. In short, this model assumes that the primary interaction among species is competition, and further, that, with few exceptions, the resources available to these species are fixed. When there are very few species in the environment, competitive interactions matter little, and the overall rate of species accumulation is exponentially increasing. As the number of species increases, competitive pressures slow down the rate of species diversification.

Sepkoski has supported his claim that the number of species in a closed system tends toward some equilibrium value by examining the fossil record for marine bivalves (1978, 1979, 1984). This group of species is one of the most likely to demonstrate the competitive use of niche space. Marine bivalves occupy sea floors. They compete for space and generally use very similar food sources by filtering their food from the water column. There are therefore two means of competitive exclusion: space limitation and food limitation.

Marine families

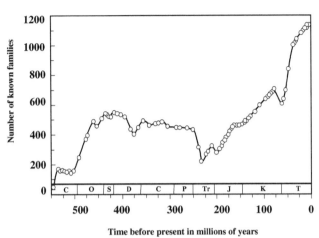

Figure 4.2. The number of marine families as represented in the fossil record (after Sepkoski 1992).

Further, there are unlikely to be any complex trophic interactions among members of this group. The species are likely to be related only through competitive interactions. In short, if species accumulation curves are driven solely by competitive interactions, this is certainly a group that should demonstrate this mechanism.

In spite of this extremely favorable system for detecting competition-limited species accumulation, the fossil data do not unequivocally support the use of a logistic model (Fig. 4.2). Clearly, the fossil record used to support Sepkoski's model predictions is not a simple rise to a clearly defined equilibrium density. Sepkoski explains deviations from the model prediction as a result of external disturbances and introduces stochastic variation into the model predictions (Sepkoski 1978). Sepkoski and co-authors make claims about the similarity of the species accumulation curves and the model predictions under conditions of random stochastic disturbances (Sepkoski 1984). The justification of the magnitude and frequency of the disturbances is based on the fossil record itself.

It is certainly not clear that the observed species accumulation curves can be well described by a logistic growth curve, even with stochastic variation. To account for some of the discrepancies, Sepkoski later fit a "tri-phasic" logistic curve to the marine bivalve data (1984). He divided the bivalves into three different groups that waxed and waned separately. The more general logistic model he proposed predicted the growth of each clade as: $dS_i/d_t = r_iS_i(1 - \Sigma S_x/K_i)$. This model assumes that the clades compete for the same resources, but that large groups of organisms, the individual clades, compete as a group against other clades. This model is sufficiently complex that only numerical solutions are possible. Sepkoski solves this set of equations by resetting all negative values to zero and using ten free parameters. Sepkoski attempts to allay readers' concern with the claim that "this may leave the impression that (the equation) could be fitted to any given situation. However, as argued below, coupled logistic equations predict specific kinds of patterns . . . which appear to be robust over paleontologically reasonable ranges of parameter values" (Sepkoski 1996, 226).

Unfortunately, one can achieve quite good agreement between data and any given model by adding disturbance of the appropriate form. More importantly, the logistic is an extremely flexible curve, which can be fit to many types of data, even data generated by a different underlying model. One is reminded of Raymond Pearle's efforts to fit successive logistic curves to human population growth (Kingsland 1985). Sepkoski answers the concern that one can fit virtually any time-series with a set of summed logistic curves by pointing out features of the fossil record which support his equilibrium interpretation, such as:

1. the explosion of new animals in the early Cambrian
2. the fast rebounds in species diversity after major extinction events, and
3. long intervals (of 200 million years or so), during which the number of species remained "stable."

It should be noted that, to date, paleontologists have rarely directly tested the assumption that species origination and extinction rates are diversity dependent. A more convincing test of the diversity-dependence model is to examine per-species (or per-taxon) rates of extinction and speciation. In his early papers, Sepkoski examined both of these rates and produced somewhat equivocal model fits. It has been quite difficult to demonstrate directly that competition occurs in natural systems, even on a paleontological scale. Demonstration of diversity or density dependence is confounded by the obscuring effects of stochastic variation. The problems with tests for density-dependent behaviors in time-series data are well understood, and are documented in the ecology literature, which, after all, was the inspiration for the modeling technique Sepkoski used. Instead of employing the more sophisticated techniques employed by ecologists (e.g., Ives 1995), paleontologists seem to rely primarily on visual assessments of curve similarity. Further, paleontologists, with very few exceptions (e.g., Benton) have not seriously considered the possibility that species accumulation curves follow other trajectories such as an exponential growth pattern, or even a random walk. The evidence used to claim that equilibrium dynamics pertain is, at best, suggestive rather than conclusive.

4.6 ALTERNATIVE MODELS AND PREDICTIONS

A glance at the species accumulation curves clearly reveals that a logistic growth in species number is not the only explanatory pattern which comes to mind. For most of the earth's history, species have been accumulating in ever-increasing numbers. An exponential increase in species number, coupled with stochastic environmental disturbances of various magnitudes, would certainly be capable of producing a similar pattern. The suggestion has been made that species accumulation curves are simply exponentially increasing, and any appearance of an approach to an equilibrium number of species is an artifact (Benton 1995).

However, Sepkoski (1996), in particular, dismisses this claim on the grounds that after large stochastic disturbances, the species number increases at a rate greater than that expected from a system with exponential growth (item 2 p. 112). He suggests that the faster rate of species accumulation is due to release from competitive pressure. At low species diversity, competitive effects will be unimportant, and species accumulation will proceed at a faster rate than near the equilibrium species diversity. However, this argument is incomprehensible. If release from competitive pressure can be expressed as the available niche space (K_s) being very large relative to the number of species, then we can say K_s tends toward infinity and $S/K_s \to 0$ for the model $dS/dt = rS(1-S/K_s)$ where dS/dt is the rate of species diversity change, r is the net species accumulation rate (origination-extinction), S is the current number of species. Therefore, the net rate of species accumulation when there is very low species diversity is simply the exponential growth rate rS.

On the other hand, major species extinctions that left surviving members of all major groups (i.e., terrestrial plants, herbivores, and predators) would be expected to recover more quickly than predicted by either the logistic model or exponential model, if the relations between these species were not strictly competitive. For example, herbivores would already exist and would merely need to diversify, rather than evolve anew, in order to provide new niche space for predators. Therefore, the species accumulation curve would rise more rapidly than if we were observing the beginning of the fossil record, where terrestrial plants, for example, did not yet exist. To use a manufacturing analogy, it takes less time to knock off a bunch of copies, as compared to coming up with an original model. Once a new technological innovation is thought out and created, copies proliferate at a very quick rate. In the same way, once a terrestrial plant has evolved, variations should appear quite quickly, even after a major extinction event that removes some of the original players.

As for Sepkoski's other lines of evidence, they are not necessarily the result of balanced speciation and extinction rates. The rapid accumulation of species after the early Cambrian is equally well explained by exponential increase in species. The periods of relatively "stable" periods are somewhat stronger evidence for logistic growth in diversity. But, stochastic variation can produce similar patterns with models of exponential, additive, cyclic, or even chaotic diversity dynamics.

Moreover, not only does Sepkoski claim that competition is the primary mechanism controlling species accumulation curves, he also suggests that only one particular kind of equilibrium behavior is a good description of these curves. That is, Sepkoski, like earlier population ecologists (Cuddington 2001), claims that not only is the logistic model a good description of the fossil species curve, but also that only a monotonic approach to a stable equilibrium is an acceptable prediction of the logistic model. Different versions of the logistic model can predict stable oscillations and chaotic dynamics that are quite different from a constant species number.

Sepkoski makes this claim by first arguing against a discrete-time version of the logistic model, like that used by Kitchell and Carr (1985) to describe the same data set. A continuous-time version of the simple logistic model has only one possible prediction: that of a monotonic approach to an equilibrium species diversity. On the other hand, a discrete-time version of this same model ($S_{t+1} = S_t r(1 - S_t/K)$), has a full range of dynamic behaviors which depend on the net rate of species origination. At high rates, more complex dynamic behavior is possible (Fig. 4.3). Sepkoski claims that the discrete-time version of the model is designed to describe periodically occurring events, and quite reasonably insists that this is not descriptive of species origination and extinction.

On this basis, Sepkoski might have eliminated the possibility that dynamic behavior other than a simple approach to a fixed equilibrium state was likely. However, to insist on instantaneous rates of species origination and speciation presents another problem. If competition is the main structuring mechanism of species diversity, and this interaction occurs without time delays, then competing species should almost never co-occur in the fossil record. The inferior competitor would be eliminated so quickly there would be little chance of preservation. Sepkoski sidesteps this problem by claiming that evolutionary

Effective Speciation Rate

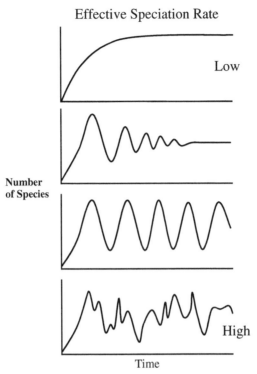

Figure 4.3. The range of diversity dynamics predicted by a discrete-time version of Sepkoski's logistic model.

innovations would allow the prolonged co-existence of competitors and increase the possibility of simultaneous preservation (1996). Sepkoski's argument is reasonable, but unfortunately undoes his previous justification for using a simple continuous-time logistic model. Although it may not be reasonable to describe species accumulation with a periodic timestep, it certainly sounds like Sepkoski is describing delayed competition effects. Any feature of the biological interaction that introduced time delays would extend the range of possible dynamic behavior predicted by a logistic growth-type model. There is no a priori reason why a simple model of species accumulation, based on a mechanism of competition with some degree of time-delayed action, should predict an approach to a single stable equilibrium.

Neither the current analysis of fossil evidence nor the arguments designed to bolster confidence in equilibrium models of species diversity is particularly convincing. Yet the large majority of paleontologists seem to accept this interpretation of the fossil record. This position is particularly suspect since the family of models chosen to support the theory that species number reaches and remains near an equilibrium number do not necessarily make this prediction. That is, paleontologists need to argue that not only does a logistic model describe species accumulation curves, but that only one prediction of this model is appropriate. Rather than subjecting this model to rigorous tests, as in population ecology, it has simply been adopted as the most feasible explanation of macroevolutionary patterns. Further, the potential role of positive-density dependence on biodiversity has been ignored.

We suggest that there is a strong basis toward equilibrium interpretations of natural systems, even among modern researchers. In a recent paper, Alroy (1999) equates Benton's rejection of the logistic model with the claim that "no coherent interpretation is possible." This bias seems quite similar to that expressed by Darwin. Namely, competition is thought to be the only important mechanism structuring species accumulation curves, and further, these curves tend toward an orderly and predictable equilibrium state.

4.7 CONCLUSIONS

Darwin, and many modern paleontologists, may have had a predisposition to interpret the fossil record and the processes of natural selection as pointing to an equilibrium biodiversity, where competition is the only important biological mechanism governing this equilibrium. This predilection probably reflects two separate influences: an image of nature as stable and benevolent, and a belief in the plenitude of natural systems. Emphasis on competition leads to the neglect of positive influence of species number on species accumulation.

Darwin and later commentators on the fossil record have emphasized the role of competition in determining species diversity, and have downplayed the role of species facilitation. The view of competition as the major biological control of species number depends on a view of natural systems as replete with species. This conception of plenitude predates modern science (Rieppel 1984). As we saw previously, although Darwin used this image of natural systems, he was also aware of systems that were clearly not niche-limited (e.g., his description of the Cape of Good Hope). Similarly, although modern paleontologists are certainly aware of positive diversity dependence, they do not explain its role in determining the fossil record.

We do not wish to suggest that Sepkoski's approach (or Darwn's!) has had a detrimental effect on paleontology. Rather, we believe that the application of a predictive model to this field has set the stage for important advances. However, we do wish to claim that this model has been received rather uncritically. There is no a priori reason to expect a stable equilibrium of species number, even if one grants that competition is the only major mechanism controlling species accumulation. Further, there is at least logical ground on which to claim that species origination may have positive density dependence. This lack of skepticism, particularly in the face of inconclusive evidence, points to the role of other influences. In particular, our Western heritage posits a predictable and bountiful nature (Egerton 1973) that may or may not exist.

BIBLIOGRAPHY

Alroy, J. 1999. Equilibrial Diversity Dynamics in North American Mammals. In *Biodiversity Dynamics: Turnover of Populations, Taxa and Communities*, M. L. McKinney and J. Drake, eds. New York: Columbia University Press.

Benton, M. J. 1987. Progress and Competition in Macroevolution. *Biological Reviews* 62:305–38.

1995. Diversification and Extinction in the History of Life. *Science* 268:52–8.

1997. Models for the Diversification of Life. *Trends in Ecology and Evolution* 12: 490–5.

1999. The History of Life: Large Databases in Palaeontology. In *Numerical Paleobiology: Computer-based Modelling and Analysis of Fossils and Their Distributions*, D. A. T. Harper, ed. New York: John Wiley & Sons.

Carr, T. and J. Kitchell. 1980. Dynamics of Taxonomic Diversity. *Paleobiology* 6: 427–43.

Cracraft, J. 1985. Biological Diversification and Its Causes. *Annals of the Missouri Botanical Garden* 72:794–822.

Cuddington, K. 2001. The 'Balance of Nature' Metaphor and Equilibrium in Population Ecology. *Biology and Philosophy* 16:437–61.

Darwin, C. 1859. *The Origin of Species*. 1st ed. London: John Murray.

1872. *The Origin of Species*. 6th ed. New York: Mentor.

Egerton, F. N. 1973. Changing Concepts of Balance of Nature. *Quarterly Review of Biology* 48:322–50.

Gale, B. 1972. Darwin and the Concept of a Struggle for Existence: A Study in the Extrascientific Origins of Scientific Ideas. *Isis* 63:321–44.

Ghiselin, M. 1995. Perspective: Darwin, Progress and Economic Principles. *Evolution* 49:1029–37.

Hoffman, A. 1985. Biotic Diversification in the Phanerozoic: Diversity Independence. *Palaeontology* 28:387–91.

Ives, A. 1995. Measuring Resilience in Stochastic Systems. *Ecological Monographs* 65:217–33.

Kingsland, S. E. 1985. *Modelling Nature: Episodes in the History of Population Ecology*. Chicago: University of Chicago Press.

Kitchell, J. A. and T. R. Carr. 1985. Nonequilibrium Model of Diversification: Faunal Turnover Dynamics. In *Phanerozoic Diversity Patterns*, J. W. Valentine, ed. Princeton, N.J.: Princeton University Press.

Miller, A. and J. J. Sepkoski. 1988. Modeling Bivalve Diversification: The Effect of Interaction on a Macroevoutionary System. *Paleobiology* 14:364–9.

Rieppel, O. 1984. The Problem of Extinction. *Zeitschrift fur Zoologischer Systematik und Evolutionsforschung* 22:81–5.

Rosenzweig, M. L. 1995. *Species Diversity in Space and Time*. Cambridge: Cambridge University Press.

Sepkoski, J. J., Jr. 1978. A Kinetic Model of Phanerozoic Taxonomic Diversity I. Analysis of Marine Orders. *Paleobiology* 4:223–51.

——— 1979. A Kinetic Model of Phanerozoic Taxonomic Diversity II. Early Phanerozoic Families and Multiple Equilibria. *Paleobiology* 5:222–51.

——— 1984. A Kinetic Model of Phanerozoic Taxonomic Diversity III. Post-paleozoic Families and Mass Extinctions. *Paleobiology* 10:246–67.

——— 1992. *A Compendium of Fossil Marine Animal Families*. 2d ed. Milwaukee Public Museum, Milwaukee, Wisconsin: Contributions in Biology and Geology, no. 83.

——— 1996. Competition in Macroevolution: The Double Wedge Revisited. In *Evolutionary Paleobiology*, D. Jablonski, D. H. Erwin, and J. H. Lipps, eds. Chicago: University of Chicago Press.

Walker, T. and J. Valentine. 1984. Equilibrium Models of Evolutionary Species Diversity and the Number of Empty Niches. *American Naturalist* 124:887–99.

5

Biological Diversity, Ecological Stability, and Downward Causation

GREGORY M. MIKKELSON

5.1 INTRODUCTION

The magnificent variety of life on Earth has astonished members of our own species ever since we arrived on the scene. Deep ecologists have honored this sense of awe by positing biological diversity as a moral end in itself. However, all of the other ethical paradigms considered by Oksanen (1997) treat biodiversity as only a means toward (and not a constituent of) "the flourishing of human and non-human life."

Nevertheless, most experts agree that whether diversity has any intrinsic value or not, it does have a myriad of key instrumental values. For example, it facilitates the "delivery" of "ecosystem services" such as carbon dioxide absorption, flood control, and nutrient cycling (Wilson and Perlman 1999). Important as their considered judgment is on this matter, scientists have only recently gotten around to testing it through experiment, theory, and systematic observation. One candidate mechanism is the contribution of species richness (number of species) to ecological stability. Presumably, more stable ecological systems provide ecosystem services more reliably.[1]

Given the ongoing wave of extinction wreaked by current forms of human economic activity, a revitalized research program on diversity–stability relations may seem to have come in the nick of time. However, it has encountered staunch resistance from certain quarters in ecology. In this chapter, I shall consider this research program, and challenges to it, in light of another debate, within philosophy and science: holism versus reductionism. In

Thanks to Peter Brown, Mario Bunge, Avi Craimer, David Davies, Genevieve Gore, James Justus, Jeff Mikkelson, Markku Oksanen, Anya Plutynski, Lisa Sideris, Rafael Ziegler; and audiences at McGill University, the University of Western Ontario, and the 2003 meeting of the International Society for the History, Philosophy, and Social Studies of Biology, for helpful comments and/or criticism.

particular, I will use the diversity–stability case to illustrate why downward causation is far more common and important than has previously been acknowledged. This fact, in turn, goes against reductionism, which downplays downward causes.

5.2 MECHANISTIC APPROACHES TO ECOLOGY: A NEW HOLISM?

Ecology, like other scientific disciplines, examines entities that collectively span several different levels of organization. Three particular levels form a perfect hierarchy of parts and wholes: organisms, populations, and communities. Each population – at the intermediate level in this three-part hierarchy – is exhaustively composed of organisms, which are at the lowest level. Each entity at the highest, community level, is, in turn, composed exhaustively of populations. With respect to such a compositional hierarchy:

1. The *reductionist* considers "upward" causal and explanatory relationships more important than "downward" causation and explanation. In other words, lower-level causes of higher-level effects are more important than higher-level causes of lower-level effects.
2. The *holist*, on the other hand, considers downward causation and explanation more important.

Campbell (1974), Wimsatt (1976), and Wilson (1988) each made important contributions to our understanding of reductionism and holism as construed here. However, their discussions leave the impression that natural selection is the only form of downward causation in the universe.

Effects of diversity (i.e., species richness – a property of entire *communities*) on the stability of component *populations* are another example of downward causation. Thus, the population-level "diversity–stability hypothesis,"[2] promoted by MacArthur (1955) and Elton (1958), is an example of downward explanation.[3] But so is the challenge to that hypothesis mounted by May (1973), Tilman (1996), and others. These latter asserted that diversity undermines population stability. Positive or negative, any effect of diversity on population stability is downward causation. Herein, I shall give reasons for expecting downward causes to pervade ecology, and perhaps all areas of scientific inquiry.

Wimsatt (1976) conjectured that most scientists are reductionists. The articles cited in the preceding paragraph may not fit this conjecture very well. One by Goodman (1975), however, does. In his highly critical review of the diversity–stability hypothesis, Goodman did not affirm May's (1973) contrary conclusion that diversity has a negative effect on population stability. Instead, Goodman simply denied any positive effect: "the diversity–stability hypothesis is a poor predictor of biological reality" (p. 251).

Experiments later confirmed May's (1973) conclusion. These same experiments (Tilman 1996) indicated a *positive* effect of diversity on *community* stability.[4] This latter type of effect is neither upward, nor downward, but same-level. Prima facie, then, it should worry neither the holist nor the reductionist. Prima facie, it should not even trouble the *strict* holist or reductionist, defined as follows:

3. A *strict reductionist* is a reductionist who believes that downward explanations are illegitimate, and should therefore be expunged altogether from science.
4. A *strict holist* is a holist with a correspondingly motivated "urge to purge" upward explanation.

I do not know of any strict holists in this sense. However, Schoener (1986) countenanced "only one path of causation in biological communities . . . lower levels (populations and/or individuals) determine (explain) entities, processes, and properties of higher levels" (Inchausti 1994, 214). Schoener thus advocated strict reductionism in ecology.

Despite the apparent innocence of same-level causes, they do pose a problem for the strict reductionist. The reason is that any higher- or lower-level "mechanism" for a same-level causal relationship must involve both downward and upward causation. To clarify this point, I propose the following explication of the "mechanism" concept used by ecologists. Given that some property A causes some other property B,

5. A *mechanism* for that causal relationship is a "pathway," that is, a set of intermediate causal steps, from A to B.

Suppose that A and B are same-level properties of a given "focal" entity. We can represent a purely lower-level mechanism for their causal relationship as $A \rightarrow L \rightarrow B$. Here, L involves only lower-level properties – in other words, properties of the parts comprising the focal entity. Obviously, then, the first step in the mechanism ($A \rightarrow L$) is a downward causal relation, while the second ($L \rightarrow B$) is upward. The same argument applies to any purely higher-level mechanism, except that the upward step comes first.[5]

The above desiderata imply that in a field like ecology, the strict reductionist must avoid many references to same-level causal relationships, as well as any to downward causes.[6] The most effective and legitimate way to eliminate such references would be to "explain away" all apparent downward and same-level causal relationships, or to otherwise furnish compelling evidence that they are spurious or nonexistent. Strict reductionism thus leads to eliminativism. For example, Huston (1997) did his best to explain diversity–stability relations as mere artifacts of experimental design.

If, however,

a. Same-level causal relationships prove robust enough to resist attempts at elimination, and important enough to figure in scientific explanations; and

b. Higher- or lower-level mechanisms for such relationships prove equally robust and important;

then the strict-reductionist program will have failed. Also, providing a plausible mechanism for a phenomenon often helps to ensconce it as robust and important.[7] This "positive feedback" from (b) to (a) further decreases the strict reductionist's chances of success.

The diversity-community-stability relation reported by Tilman (1996) may or may not prove to be sufficiently robust.[8] Lehman and Tilman (2000) did propose three plausible mechanisms for this positive relationship. We can schematize these mechanisms as follows:

(i) $D \rightarrow$ Changes in resource-competitive population dynamics $\rightarrow O, P,$ and/or $C \rightarrow S$

(ii) $D \rightarrow$ Changes in Lotka-Volterra-like population dynamics $\rightarrow O, P,$ and/or $C \rightarrow S$

(iii) $D \rightarrow$ Changes in broken-stick-like population dynamics $\rightarrow O, P,$ and/or $C \rightarrow S$

Here, D stands for diversity, and "$\rightarrow$" means "causes." The three types of population dynamics are based on three well-known modeling paradigms in ecology. $O, P,$ and C represent what Lehman and Tilman called "the overyielding effect," "the portfolio effect," and "the covariance effect," respectively. Overyielding is an increase in mean community biomass.[9] The portfolio effect is a decrease in the sum of the variances of the individual species' biomasses. And the covariance effect is an enhanced level of biomass "compensation" among competing species. That is, if the biomass of one population increases, the biomass of one or more other population(s) tends to decrease, and vice versa. Finally, S represents an increase in community stability.

All three of these mechanisms (when rendered mathematically explicit) generate positive diversity-community-stability relations. This means that those relations are robust in at least one sense (cf. Levins 1966). Also, as suggested previously, the first step in each mechanism – from (community) diversity to population dynamics – involves downward causation. These developments do not bode well for strict reductionism.

In all three of the mathematical models of diversity–stability relations proposed by Lehman and Tilman (2000), interactions between populations play an important role.[10] Believe it or not, one prominent school of thought purports to treat populations as if they do not interact at all. In many cases, however, members of this "individualistic" or "autecological" school make assumptions that entail interspecific interaction. Nevertheless, at least one truly individualistic challenge to a prevailing "interactionist" construal has enriched the understanding of diversity–stability relations.[11]

Hagen (1989) identified three main schools of ecological thought. One centers on phenomena such as the flow of energy through entire ecosystems.[12] The other two focus on properties of the populations contained within ecosystems.[13] The first of these – what I shall call the interactionist approach – highlights competitive, predatory, and/or mutualistic interactions between populations.[14] Practitioners of the other "individualistic" approach, on the other hand, often imply that such interactions are unimportant. As Gleason (1926, 26) put it, "every species . . . is a law unto itself."

Simberloff (1982) called the "individualistic" approach "population-reductionist," and held out the hope that it could explain away many patterns found at the community level.[15] However, "individualism" often fails to live up to its name. Often, it merely hides, rather than truly eliminates, top-down constraints on, and interactions between, individuals. For example, the physicists who achieved the so-called "atomistic reduction" of thermodynamics assumed the top-down constraint that energy remains constant at the level of the whole system. This suggests that the energy levels of individual gas particles are not independent of each other (Garfinkel 1981). In political theory, Garfinkel accused Nozick (1974) of pretending that the economic holdings of individual people in a society are mutually independent.

"Individualistic" ecologists appear to have made similar mistakes. For example, Gleason (1939, 94) did not acknowledge the full significance of a top-down constraint that he himself affirmed: "The land surface of the world is already fully occupied by plants. Room for additional plants is made available only by the death of plants now existing." Call this observation the "Gleasonian provision." This provision entails strong intra- and interspecific competition for space, and thus contravenes Gleason's own claim that species are independent of each other.[16] Hubbell (2001) built the Gleasonian provision into his "unified neutral theory of biodiversity and biogeography." Unfortunately, Hubbell, too, underplayed the provision's implication of stiff competition between species (Mikkelson in review).[17]

Despite these conceptual problems, individualistic ecology has contributed at least one important insight about diversity and stability. Tilman (1996) originally interpreted his experimental results in terms of the covariance effect – that is, negative (competitive) covariance between species. However, Doak et al. (1998) showed that diversity would enhance community stability even if populations did not interact at all. Their paper focused on the portfolio effect, which, like the covariance effect, does involve interaction between species. However, in one of their models, population biomasses and variances are completely insensitive to (i.e., undiminished by) increased diversity; and population biomasses vary independently of each other. Under these conditions, overyielding ensures that community stability increases with diversity.

5.4 HIGHER- VERSUS LOWER-LEVEL LAWS

In previous sections, I offered several arguments against strict reductionism. In this section, I shall challenge one potential reason for adopting milder versions of reductionism: that lower-level generalizations are more "lawlike," or robust, than higher-level generalizations.[18] In ecology, the opposite seems to be true, at least with respect to the traditional hierarchy of individuals, populations, and communities.

In Chapter 3 of the *Origin* ("Struggle for Existence"), Darwin (1859) cited example after example of bewildering ecological complexity – at the organismal and population levels. Such "ever-increasing circles of complexity," he admitted, often tempt naturalists to attribute population phenomena to "what we call chance" rather than "definite laws." In contrast, Darwin asserted a rather simple community-level pattern in Chapter 4 ("Natural Selection"). Diversity, he said, enhances overall community biomass. In other words, Darwin attested to the overyielding effect.

Increasingly, ecologists (e.g., Brown et al. 2001 and Ernest and Brown 2001) are finding community properties to be more stable than population properties. I hypothesize that this contrast in stability generates another contrast: in the "lawlikeness" of relationships involving variables found at the two levels. As one – though certainly not the only – conceivable measure of lawlikeness, I propose the strength of correlations between independent and dependent variables.[19]

For example, consider the experiments that helped to re-ignite interest in diversity–stability relations (Tilman 1996).[20] These experiments were originally designed to test the effects, not of diversity, but of nitrogen fertilizer.[21] "Response" (i.e., dependent) variables included community and population biomasses.[22] How tightly does nitrogen "control" these variables? Does it exert more of an effect at the population or community level? As it turns out, community biomass is quite a bit more tightly correlated with nitrogen than are population biomasses.[23] Diversity, another community variable, correlates more strongly yet with nitrogen.[24]

Does the same contrast exist when population and community properties are considered as independent, rather than dependent, variables? With respect to many of the ecosystem services mentioned previously, this does seem likely. For example, all plants take in carbon dioxide during photosynthesis. Presumably, then, total community biomass will correlate better with CO_2 absorption than will the biomass of any particular population. Diversity, too, may explain more variation in productivity, and thus CO_2 uptake, than do individual population biomasses (Tilman et al. 2001).

5.5 CONCLUSION

In the 1990s, scientists responded to alarm about progressive damage to the Tree of Life by embarking on a renewed program of research into the relationships between biological diversity and ecological stability. This program has helped to advance and unify the field of ecology, by elucidating relationships between various levels of organization. In this chapter, I have focused on one implication of these developments: that downward causation plays more of a role in nature than has been recognized heretofore by scientists, and even by philosophers.

In particular, I have argued that:

1. Natural selection and diversity-population-stability relations are two examples of downward causation.
2. Any lower-level (or higher-level) mechanism for a same-level causal relationship must involve downward causation.

    a. Diversity-community-stability relations are same-level causal relationships.

    b.   Several plausible, lower-level mechanisms have been proposed for diversity-community-stability relations.

    c.   Therefore, these mechanisms illustrate one way in which references to downward causal relationships ought to proliferate in science.

3.   Although individualists frequently deny the importance of interactions between populations, they often make assumptions, such as the Gleasonian provision, that entail strong top-down constraints. These constraints, in turn, entail strong interactions between populations.

4.   In ecology, higher-level generalizations tend to be more lawlike than lower-level generalizations.

These arguments imply that strict reductionism – an exclusive emphasis on same-level and upward causation – is untenable. What about a milder form of reductionism – the view that upward causes are simply more important than downward causes? Most of these considerations suggest that neither type of cause deserves predominant emphasis.[25] However, the greater lawlikeness of higher-level generalizations in ecology may suggest that downward causes should be favored within that field.

Should this turn out to be true in other fields as well, it would call for a substantial re-orientation of scientific funding practices. Odum (1977) put it this way:

> This is not to say that we abandon reductionist science, since a great deal of good for mankind has resulted from this approach, and some of our current short-range problems can perhaps be solved by this approach alone. Rather, the time has come to give equal time, and equal research and development funding, to the higher levels of biological organization . . . It is . . . the properties of the large-scale, integrated systems that hold solutions to most of the long-range problems of society.

Twenty-six years later, the case for a sea change in scientific funding is even more compelling. This means relatively less money for such reductionistically-driven ventures as the Human Genome Project, and much more for holistically inspired endeavors such as what we might call the "Earth Speciome Project" (Handwerk 2002).

NOTES

1.   A person inspired by Leopold's (1949) land ethic might also view stability as an end in itself.

2.   In other words, that diversity enhances population stability.

3. By a "downward explanation," I mean an explanation that emphasizes downward causation.

4. The relevant definition of stability here is the mean, divided by the standard deviation, of biomass (Lehman and Tilman 2000). Biomass is the sheer amount of living material contained within an organism, population, or community.

5. See Darden (2000) for a potential example of a purely higher-level mechanism.

6. By "a field like ecology" I mean one in which many practitioners are keen to elucidate lower-level mechanisms for higher-level phenomena.

7. For example, the mechanism of natural selection helped Darwin (1859) to establish the fact of evolution, and various plate-tectonic mechanisms helped to establish the fact of continental drift (Sterelny and Griffiths 1999).

8. One earlier study (McNaughton 1993) also indicated a positive relationship between diversity and community stability, as did five out of seven other papers reviewed by Cottingham et al. (2001).

9. This pattern has also been challenged (Huston 1997).

10. This includes their simplest, broken-stick-type model. Its assumption of a fixed upper limit on community biomass entails interspecific competition.

11. See Naeem (2002) for more on the fertility of debates about the ecological effects of biological diversity.

12. Hagen saddled this approach with Hutchinson's (1978) awkward term "holological."

13. These approaches are both types of "merology," in Hutchinson's and Hagen's terminology.

14. Hagen referred to interactionism as the "equilibrium" approach. However, the research tradition to which he referred has also incorporated nonequilibrium thinking.

15. "Community level attributes . . . are primarily epiphenomena of component population attributes" (p. 142).

16. See Garfinkel (1981) on how the "Lockean proviso" likewise renders many of Locke's own arguments inapplicable.

17. Yu et al. (1998) made a similar point about Hubbell's theory.

18. Inchausti (1994, 211) ascribed a related view to reductionists: "theories regarding lower-level phenomena are 'more general' and 'more explanatory' than higher-level ones."

19. See Mikkelson (in press) for another application of this idea.

20. Data posted at www.lter.umn.edu/research/research.html, as part of Experiment E001.

21. Tilman (1996) used statistical techniques to rule out certain effects of nitrogen that might otherwise have confounded the relationship between diversity and stability. Nevertheless, Huston (1997) proposed at least one additional confounding effect. Tilman et al. (1994) had admitted that "a cleaner test of the effects of biodiversity . . . would come from direct experimental control of species diversity. In 1993 we began such a study." Unfortunately, they have not yet reported any effects of these direct diversity manipulations on stability.

22. Effects of nitrogen and other abiotic factors on community properties, and vice versa, are not clearly downward or upward relationships. A reductionist might consider nitrogen – a physical variable – to be at a lower level than community

biomass. A holist, however, might counter that nitrogen levels are properties of ecosystems, which are often considered to be at a higher level than communities. Such ambiguities help to demonstrate the value of limiting questions about reductionism and holism to strict part-whole hierarchies.

23. The absolute value of the correlation (ρ) between community biomass and nitrogen is .35. The mean (weighted by mean population biomass) of the absolute values, of the correlations between population biomasses and nitrogen, is .21.
24. $|\rho| = .42$.
25. Bunge (2000) called something like this balanced view "systemism."

BIBLIOGRAPHY

Brown, J. H., S. K. M. Ernest, J. M. Parody, and J. P. Haskell. 2001. Regulation of Diversity: Maintenance of Species Richness in Changing Environments. *Oecologia* 126:321–32.

Bunge, M. 2000. Systemism: The alternative to Individualism and Holism. *Journal of Socio-Economics* 29:147–57.

Campbell, D. T. 1974. "Downward Causation" in Hierarchically Organised Biological Systems. In F. J. Ayala and T. Dobzhansky, eds. *Studies in the Philosophy of Biology*. Berkeley: University of California Press, pp. 179–86.

Cottingham, K. L., B. L. Brown, and J. T. Lennon. 2001. Biodiversity May Regulate the Temporal Variability of Ecological Systems. *Ecology Letters* 4:72–85.

Darden, L. 2000. How Scientists Explain Disease. *Philosophy of Science* 67:352–4.

Darwin, C. 1859. *On the Origin of Species by Means of Natural Selection*. London: John Murray.

Doak, D. F., D. Bigger, E. K. Harding, M. A. Marvier, R. E. O'Malley, and D. Thomson. 1998. The Statistical Inevitability of Stability–Diversity Relationships in Community Ecology. *The American Naturalist* 151:264–76.

Elton, C. S. 1958. *The Ecology of Invasions by Plants and Animals*. New York: Chapman and Hall.

Ernest, S. K. M. and J. H. Brown. 2001. Homeostasis and Compensation: The Role of Species and Resources in Ecosystem Stability. *Ecology* 82:2118–32.

Garfinkel, A. 1981. *Forms of Explanation*. New Haven, CT: Yale University Press.

Gleason, H. A. 1926. The Individualistic Concept of the Plant Association. *Bulletin of the Torrey Botanical Club* 53:7–26.

Gleason, H. A. 1939. The Individualistic Concept of the Plant Association. *American Midland Naturalist* 21:92–110.

Goodman, D. 1975. The Theory of Diversity–Stability Relationships in Ecology. *Quarterly Review of Biology* 50:237–66.

Hagen, J. B. 1989. Research Perspectives and the Anomalous Status of Modern Ecology. *Biology and Philosophy* 4:433–55.

Handwerk, B. 2002. Team Races to Catalog Every Species on Earth. *National Geographic News*. March 5.

Hubbell, S. P. 2001. *The Unified Neutral Theory of Biodiversity and Biogeography*. Princeton, NJ: Princeton University Press.

Huston, M. A. 1997. Hidden Treatments in Ecological Experiments: Re-evaluating the Ecosystem Function of Biodiversity. *Oecologia* 110:449–60.

Hutchinson, G. E. 1978. *An Introduction to Population Ecology*. New Haven, CT: Yale University Press.

Inchausti, P. 1994. Reductionist Approaches in Community Ecology. *The American Naturalist* 143:201–21.

Lehman, C. L. and D. Tilman. 2000. Biodiversity, Stability, and Productivity in Competitive Communities. *The American Naturalist* 156:534–52.

Leopold, A. 1949. *A Sand County Almanac*. New York: Oxford University Press.

Levins, R. 1966. The Strategy of Model Building in Population Biology. *American Scientist* 54:421–31.

MacArthur, R. H. 1955. Fluctuations of Animal Populations, and a Measure of Community Stability. *Ecology* 36:533–6.

May, R. M. 1973. *Stability and Complexity in Model Ecosystems*. Princeton, NJ: Princeton University Press.

McNaughton, S. J. 1993. Biodiversity and Function of Grazing Ecosystems. In E.-D. Schulze and H. A. Mooney, eds. *Biodiversity and Ecosystem Function*. New York: Springer-Verlag, pp. 361–83.

Mikkelson, G. M. In press. Ecological Kinds and Ecological Laws. *Philosophy of Science*.

In press. Ecosystems as Organisms. In R. A. Skipper, Jr., R. Ankeny, C. F. Craver, L. Darden, G. M. Mikkelson, and R. C. Richardson, editors. *Philosophy and the Life Sciences: A Reader*.

Naeem, S. 2002. Ecosystem Consequences of Biodiversity Loss: The Evolution of a Paradigm. *Ecology* 83:1537–52.

Nozick, R. 1974. *Anarchy, State, and Utopia*. New York: Basic Books.

Odum, E. P. 1977. The Emergence of Ecology As a New Integrative Discipline. *Science* 195:1289–93.

Oksanen, M. 1997. The Moral Value of Biodiversity. *Ambio* 26:541–5.

Schoener, T. W. 1986. Mechanistic Approaches to Ecology: A New Reductionism? *American Zoologist* 26:81–106.

Simberloff, D. S. 1982. Reply. In E. Saarinen, ed. *Conceptual Issues in Ecology*. Boston: D. Reidel, pp. 139–54.

Sterelny, K. and P. Griffiths. 1999. *Sex and Death: An Introduction to Philosophy of Biology*. Chicago: University of Chicago Press.

Tilman, D. 1996. Biodiversity: Population versus Ecosystem Stability. *Ecology* 77:350–63.

Tilman, D., J. A. Downing, and D. A. Wedin. 1994. Reply. *Nature* 371:114.

Tilman, D., P. B. Reich, J. Knops, D. Wedin, T. Mielke, and C. Lehman. 2001. Diversity and Productivity in a Long-term Grassland Experiment. *Science* 294:843–5.

Wilson, D. S. 1988. Holism and reductionism in evolutionary ecology. *Oikos* 53:269–73.

Wilson, E. O. and D. L. Perlman. 1999. *Conserving Earth's Biodiversity*. Washington, DC: Island Press.

Wimsatt, W. C. 1976. Reductive Explanation: A Functional Account. In R. S. Cohen, ed. *PSA 1974*, pp. 671–710.

Yu, D. W., J. W. Terborgh, and M. D. Potts. 1998. Can High Tree Species Richness Be Explained by Hubbell's Null Model? *Ecology Letters* 1:193–9.

Part III

Valuing Biodiversity

6

Jean-Jacques Rousseau

Philosopher as Botanist

FINN ARLER

Jean-Jacques Rousseau is best known as a philosopher, if not a writer of fiction. And yet, in the last decade of his life he devoted much of his time to empirical studies of biological organisms. These studies seem to have been quite extensive. If we confine ourselves to writers who are lined up in traditional histories of philosophy, Rousseau's biological studies are probably outmatched only by those of Aristotle and Theophrastus. He made his own herbaria, and he even wrote a few botanical works: a (fragmentary) dictionary of botanical terms, and a series of "elementary" letters on botany to his "dear friend" Madame Delessert, who asked for advice on how to introduce her daughter Marguerite-Madeleine to the study of plants. These letters were written between 1771 and 1773, and published for the first time in Geneva three years after his death in 1778. The most famous edition, from 1805, included illustrations made by the most celebrated of all botanical illustrators, Pierre-Joseph Redouté, "the Raphael of flower painting," protégé of both Marie-Antoinette and, in particular, Josephine Napoléon, who spent a significant part of her fortune on this sponsorship.

Unlike the biological studies of Aristotle and Theophrastus, which were not superseded for two millennia, there is nothing original in Rousseau's empirical studies. He followed the sexual system of Linnaeus rather slavishly, and he does not seem to have found any new species or any features of plants which were not described before. He was well informed about the state of the art of botany, and it is obvious that he was quite observant and systematic in his studies, but he started out so late in life, and was busy with so many other activities, that he was never able to add anything significant to the science. From a modern point of view, the most interesting thing about Rousseau is not the results of his empirical studies, but rather his meditations on the goals, reasons, and values which are involved in or emerge from these studies. In the two botanical papers as well as in his late autobiographical writings, one can find various reflections on the value of or the values related to the study of biological diversity.

Before discussing his botanical studies, however, I shall take a look at a couple of other descriptions of natural phenomena, or rather, reflections on impressions of natural phenomena, which can be found in some of his other works. These descriptions or reflections can give us a hint about why botany was such an attractive field of study for Rousseau. In the first section I shall analyze a couple of his descriptions of gardens, in the second section I shall look at the way he describes a trip to the Swiss mountains.

6.1   GARDENS

It is well known that Rousseau found a symbol of the misery of his age in the contemporary baroque gardens. In the first paragraphs of *Émile* he ridicules the mistreatment of trees and shrubs in these gardens, the perverted love of monstrosities, and the excessive desire to twist and distort the natural order. Everything in these gardens is planned to be as artificial as possible, nothing is allowed to follow its own natural course. Why is that? he asks, and answers that the gardens are used as weapons in the social struggle for recognition, a struggle which everybody is bound to lose, because every individual wants more love and respect from other people than he or she would ever be willing to give him- or herself.

Thus, the gardens had become showrooms with a variety of fancy installations, including a broad array of exotic plants, and sometimes birds and beasts, imported from all the four quarters of the world. Biodiversity had become yet another means in the ongoing social struggle, wherefore the interest in the biological world was limited to the showiest organisms. The owners of gardens were trying to outrival each other by presenting new and still more strange and exotic plants and birds, together with all the other extravagant features – features which were all doomed to be superseded by new contraptions and fabrications before the end of the season.

The baroque garden included all those elements which Rousseau despised in contemporary France (although he could not deny his own attraction at the same time): falseness, vanity, artificiality, exaggeration. But, then, which kind of garden would he truly prefer instead? There are actually several possibilities (cf. Arler 1997), but I shall confine myself in this chapter to the two descriptions of gardens that tell us most explicitly about his own ideas of the ideal kinds of gardens, and indirectly about his views on the value of biological diversity. The first one is the description of Sophy's parents' garden in

134

book V of *Émile*. The second one describes Julie's Elysium in the fourth part of *Julie or the New Heloïse*.

Rousseau's description of Sophy's parents' garden is to a large extent based on Homer's story of Ulysses's stay in king Alcinous's garden in Book VII of *The Odyssey*, a garden that is cool, pleasant, and simple, but still filled with a variety of plants, all of which are both useful and beautiful: apple and pear trees, figs, vines, herbs, and olive trees. Sophy's parents' garden is organized the same way (Rousseau 1762/1993, 458f.; cf. also 379f.). Unlike the baroque gardens, there are no gaudy flower beds, but only a well-kept vegetable garden embellished with a few local flowering plants. There is no showy park with trimmed parterres and exotic trees, but only a fruit garden with many different local fruit trees. There are no ornamental lakes or canals but only small, free-flowing streams with herbaceous banks running through the garden. A modest kind of beauty and a simple kind of usefulness are combined in every single part of the garden. This is not a garden intended for display as part of the permanent social struggle for honor and recognition.

It is interesting to note that both Homer and Rousseau focus on humans' physical need for and attraction to plants. The human body is the measure of the garden. The scale is human, and everything is arranged to satisfy human needs of two kinds. On the one hand, the owners and visitors make use of the plants. They eat and drink the products of the garden; they need the fruits and vegetables for nourishment. On the other hand, they are attracted in a physical sense not only to the good taste of the fruits and the wine, but also to the beauty of the flowers and the quiet purl from the water courses, and they become calm in surroundings which are attractive in consequence of the specific collection of plants. Both gardens have pleasant atmospheres that visitors can sense immediately. The biological diversity is limited, however: only the most useful and pleasant plants are present, whereas the rest of the plants are left in their natural habitats beyond the boundaries of the garden.

Now, let us take a look at another garden, Julie's Elysium in the philosophical novel *Julie or the New Heloïse*, a garden that is organized for very different reasons (Rousseau 1761/1997, 387ff.). Julie's garden is described in a letter written to a friend by Julie's Platonic lover, the anonymous philosopher sometimes referred to as St. Preux (the heroic saint). When he visits the garden it is the first time he has seen Julie in seven years. Her father has forced her to marry a friend of his, Monsieur de Wolmar, who is a kind and pleasant man, but whom she does not and cannot love because of her secret, Platonic communion with the philosopher. Julie calls the garden her Elysium, because it is meant to be the meeting place for souls who have left all their earthly needs and desires behind. It is hidden by a shaded avenue and heavy foliage from a variety of trees, shrubs, and twiners. There is a wall or a fence surrounding it, the gate is locked, and only very few people are allowed to enter.

Julie's Elysium is a paradise in the sense of the Old Persian word "peiri-dae 'za," a walled and protected space, as well as in all of the three senses we find in the Western tradition. First, it reminds us of the place where, allegedly, humans were once living in accord with nature, the kind of place that existed before vain ambitions made people see their relation to nature (as well as to other people) as one continuous struggle. Second, it is in itself an anticipation or representation of the utopian ideal of a coming resurrection of the long-lost harmony with the rest of nature. It is a representation of an appropriate communion with our fellow species, where humans play as modest a role as possible. Finally, like in the old myth of the Elysian fields, it is the place where souls meet when they have left their bodily cravings as well as other kinds of earthly desires.

The Elysium is a meeting ground of Platonic or Christian souls, who have no motivation apart from the transcendental Good or universal charity. In this sense it is a monument of Julie's pure and unearthly love to the philosopher, or a reminder of the afterlife in a purer world inhabited by souls without bodily needs and without the urge always to separate between mine and thine. It is a garden for purified souls, who are able to sense, but whose sensations are eccentric in the sense that their own self-centered needs and interests are put aside. In the last resort, the garden is incomprehensible unless one can contemplate it in purely intellectual terms.

St. Preux describes the garden as a wild and solitary place, and by entering he feels as if he was "the first mortal who ever had set foot in this wilderness" (Rousseau, 1761/1997, 387). It is as if he has returned to Paradise before the introduction and subsequent fall of human beings. He therefore imagines that the appearance of the garden must be the work of an uninhabited nature left on its own. Julie corrects him, however: it is not a true wilderness, it is an "artificial wilderness" (389). "It is true," she explains, "that nature did it all, but under my direction, and there is nothing here that I have not designed" (388). She designed the Elysium herself, but she did it in accordance with nature as it would have been without the presence of humans. There is not even the slightest trace of cultivation, and great care has been taken to erase all human footprints (393).

Why does the garden have to be an artificial wilderness? The reason is that the garden is intended as homage to virtue, not to innocence. To be virtuous is to do what one considers to be right *in spite of* an inclination to do the opposite. The Elysium is not for innocent people who have no real freedom and only act in accordance with their natural inclination, like beasts. It is for free spirits who are able to cool their heated passions and act as the purified heart with its sense of real duty tells them to do. In a wilderness there would be no virtue; only an artificial wilderness can satisfy the demands of freedom.

Everything in the garden is therefore ordered in accordance with the demands of a nature that does not include the selfish interests of human beings. This ambition can also be seen, for instance, in the owners' relation to the birds. There is an aviary in the garden, but it is not like other aviaries: the birds are not closed in by latticework. The birds are not kept for pleasure. They are free to fly away, but nevertheless prefer to stay. "I see," the philosopher says, "you want guests and not prisoners." But even this is not an appropriate description, he is taught: "Who are you calling guests?" Julie responds, "It is we who are theirs. They are the masters here, and we pay them tribute so they will put up with us occasionally" (391). Only by following a nature in which their own self-centered desires have no part can humans enjoy its pleasures in the right, eccentric way.

Unlike the gardens of Alcinous and Sophy's parents, Julie's garden is thus laid out without even the slightest thought about utility. The garden is indeed pleasant and agreeable, but this is only a secondary gain. The main purpose for Julie is to create a place where souls can meet beyond social and bodily needs, desires – and inequalities. The garden is isolated from ordinary social life, and it was never meant simply to please the senses. It is almost as if the visitors to the garden have deposited their bodies at the entrance, in order to be able to see the organisms unfold and develop without a second thought for their own previous social and bodily interests. If one can still talk about an interest in the biological diversity of the garden, it is an interest that is thoroughly purified, a disinterested interest.

6.2 MOUNTAINS

Julie or the New Heloïse includes another description of an encounter with natural phenomena that adds yet another piece to the picture Rousseau draws of our relation to biological diversity. I am thinking of the anonymous philosopher's description of his travels to the Swiss mountains. St. Preux's main reason for the trip is to put his heartaches at a distance for a while. He is beginning to realize that his love for Julie may never be satisfied in an earthly sense. It is likely that their communion will never be official, and he will probably never have the chance to unite with her in a physical way. He is also beginning to realize that Julie's love depends on his ability to transcend his own social and bodily inclinations. He must be worthy of her love, and to obtain this his own love must be as pure and unselfish as she believes her love to be. To preserve this love he must get away from her for a while, to cool the bodily passions and social ambitions in order to avoid being carried away and acting rashly. So he heads for the mountains.

Mountains are attractive to people with a philosophical temperament for obvious reasons. In the mountains everyday life is kept at a distance. One is out of reach, leaving the everyday world behind. The thin air together with the effort needed to reach the top contributes to this cleansing process. The ascent is a kind of cathartic process, which at least temporarily changes one's attitude toward the world. The very physical scale of the mountains – together with the signs of its incomprehensible time scales – add to this sense of outer-worldliness. Finally, a mountain peak is the best spot from which to view the surrounding world.

Rousseau lets the anonymous philosopher use at least some of these points in his letters to Julie. St. Preux carefully describes how he, after a long cathartic stroll through the clouds, finally reaches a site with a clear view where, in season, he could have seen thunder and storms gather below. He admits that this can all too easily be compared with the "vain image of the wise man's soul, the original of which never existed, or exists only in the very places that have furnished the emblem" (Rousseau, 1761/1997, 64). But even this sober-minded reflection cannot prevent the change of attitude which the mountains bring about. "It seems that by rising above the habitation of men one leaves all base and earthly sentiments behind, and in proportion as one approaches ethereal spaces the soul contracts something of their inalterable purity. There, one is grave without melancholy, peaceful without indolence, content to be and to think: all excessively vivid desires are blunted; they lose that sharp point that makes them painful" (64). Thoughts become purer and clearer high in the mountains, and become liberated from the disturbing sentiments that stick to them down in the valley below the clouds.

At the same time, however, St. Preux underlines yet another important experience: the almost insistent presence of natural phenomena makes him forget all his self-centered worries. He does not even have time for daydream-ing, because unexpected vistas of stones and cliffs, waterfalls and streams, dark woods and open meadows, unknown animals and strange plants show up and distract him at every turn of the trail (63). The visit to the mountain not only frees him from the sentiments that have absorbed him, and made his thoughts fuzzy. It also allows him to become absorbed in the natural phenom-ena which surround him: "all in all, the spectacle has something indescribably magical, supernatural about it that ravishes the spirit and the senses; you forget everything, even yourself, and do not even know where you are" (65).

There is an important difference between the two kinds of absorption. The absorption in sentiments has a self-centered side to it. The philosopher's love for Julie is not as pure as he wishes it to be. It is combined with an exclusive kind of longing: he does not only want the best for her, he also wants to be the one she chooses to be with. He wants satisfaction of his own needs. On the other hand, his absorption in the natural phenomena is not combined with any personal interest and ambition. He is simply fascinated by them, in a disinterested way which does not involve any self-centered sentiments. If he tries to get closer to a squirrel or a hawk, it is not because he has some exclusive or external interest, and he will not be harmed personally if they turn their back on him and disappear. If Julie did the same thing, without any satisfactory explanation, it would have devastating consequences. His absorption in self-centered sentiments makes him forget his sober-minded and impersonal reasoning capacity, whereas his absorption in the natural phenomena makes him forget the self-centered sentiments.

Rousseau unintentionally draws an interesting parallel between the philosopher's interest in the natural phenomena and the behavior of the mountain peasants toward strangers. What strikes the philosopher about the peasants is "their disinterested humanity, and their zeal of hospitality toward all strangers whom chance or curiosity leads to them" (65). Strangers are not used to gain a profit from their visit. They are welcomed and treated with extensive hospitality. This is particularly remarkable, because the peasants farther down in the valley were known to be quite rough on travellers.

The difference is explained to the philosopher by one of the mountain peasants: "In the valley, he said, the strangers who pass through are merchants, and other people solely occupied by their trade and gain. It is just that they leave us a part of their profit and we treat them as they treat others. But here where no business attracts strangers, we are sure that their travels are disinterested; and so is the welcome we reserve for them. They are guests who come to us to see us because they like us, and we greet them with friendship" (65). This friendship is not comprehensive, intimate, and exclusive, like the ones between Julie and her cousin Claire and between St. Preux and Lord Edward, but it is still generous, open-minded, and free from ulterior motives. The peasants and the visitors are mutually moved by a disinterested interest which, apart from the mutuality, is not very different from the one that leads the philosopher in his study of hawks and squirrels: this is likewise a study of pure curiosity, free from ulterior interests.

What is it that makes it possible for the anonymous philosopher to be totally absorbed in the mountains? What is it that he finds so fascinating that he forgets all his previous worries? When he really tries to convince us that he has found a heavenly place on Earth, a place worthy of his absorption, this is what he writes: "Imagine the variety, the grandeur, the beauty of a thousand stunning vistas; the pleasure of seeing all around one nothing but entirely new objects, strange birds, bizarre and unknown plants, of observing in a way an altogether different nature, and finding oneself in a new world. All that makes up an inexpressible mixture for the eye" (65). This, he notices, is the best and most instant medicine one can ever come across when trying to heal a broken heart: already during the first day "I attributed the calm I felt returning within me to the delights of variety" (63).

There is beauty and grandeur in the mountains, but the part of the answer that he keeps repeating is variety. Variety is the basic ingredient in his medicine for a broken heart, and variety is what he finds to be the most prominent feature of the Alpine landscape: "nature seemed to take pleasure in striking an opposition to herself, so different did one find her in the same place at various angles. To the east spring flowers, to the south autumn fruits, to the north the winter ice: she combined all the seasons in the same instant, every climate in the same place, contrary terrains on the same soil, and composed a harmony unknown elsewhere of the products of the plains and of the Alps" (63). Almost everything can be found there, not only in striking combinations of opposites like the one between the "grand and superb objects" and "the charms of a cheerful and rural place," but also in the constant changes in the mixture of plants and animals, which one observes on the way toward the summit (425). There is always more to see, one will never get bored from looking at the same old things day in and day out. Still, unlike the baroque gardens, nothing is arranged to arouse the curiosity for ulterior reasons.

6.3 COLLECTING PLANTS AND STUDYING BOTANY

So far I have only discussed works which Rousseau produced before he threw himself into the botanical studies. It is quite obvious that he was aware, at least intellectually, of the possible pleasures to be derived from the study of botany, years before he became preoccupied with it himself. For instance, he makes it clear to us how eager Émile's teacher is to take his pupil out on long walks where they can study stones and plants in real-life surroundings. Likewise, he brings to St. Preux's attention the many unknown plants in the mountains, and enables him to identify a series of plants when he visits Julie's Elysium.

It was not until the last one or two decades of his life, however, that Rousseau finally found the time to pursue a thorough and still more passionate study of botany.

On the seventh of the nine reflective walks narrated in *The Reveries of the Solitary Walker*, he tries to figure out why he has found the study of botany so rewarding. The first thing that comes to his mind is the same thought he has already let St. Preux point out in *Julie*: it distracts his attention from all the worries he is facing in the social world. The study of botany is a refuge where he can hide himself and at the same time collect new energy from the engagement in a pleasant diversion: "having arrived in places where I see no trace of men, I breathe more at my ease, as though I were in a refuge where their hatred no longer pursues me" (Rousseau 1782/2000, 65). He even sees it as a sweet revenge on all his (real or imagined) persecutors and tormentors, who undoubtedly will be vexed at the continuous happiness he has found in the study of plants (58).

If there were nothing truly rewarding in the study of plants, however, there would be no sweet revenge. The refugee from the social world would just be deceiving himself, if he kept insisting that the study of botany could fill the void which the withdrawal from social life has left within him. So there must be something in it that can truly compensate for the losses in the social sphere. An important part of the attraction is that it is more pleasant and less demanding than the philosophical reflections on which he has spent most of his time (58). A philosophical meditation is more laborious, tiring, and sometimes even depressing. It is not immediately rewarding the way the study of plants is. Botany is less tiresome because it can be done without continuous reflection, and, in contrast to philosophy, it includes much sensuality. Consequently, one can lose oneself easily, without being disturbed by depressing meditations and self-examinations, in the joy of the harmony and beauty of all the living wonders of nature. Just like trips to mountain peaks, botany allows a great escape from everyday life.

In this sense, the study of botany is for Rousseau very much in line with his reveries about reverting the fall of man, trying to become truly united with Mother Nature once again. When one loses oneself in the study of one's fellow creatures, it is almost like returning to a state of nature, where there existed immediacy and transport of delight, and where no vanity, ulterior motives, or social quarrel and competition were degrading human life. In the study of nature one can still regain some of the experiences of true belonging and selfless affection, which seem so rare in contemporary social life. "Oh, Nature! Oh, my mother!" he bursts out passionately, when he is finally alone on the little island Saint-Pierre, on which the citizens of Bern had allowed him to settle, "behold me under thy protection alone! Here there is no cunning or knavish mortal to thrust himself between me and thee" (Rousseau 1789/1996, 632). Not only is there no need for the philosophical meditations, which always cut the immediate bond between humans and nature. The disinterested study of nature can even restore at least some of the *joie de vivre* which a fraudulent and deceitful social life had almost made us forget.

The wish to lose oneself in nature has a double meaning here. On the one hand, Rousseau has a deep longing to lose his status as a free but therefore also separate individual. He wants to be a tiny part of a big continuous world order, or to return to the arms or womb of the big Mother Nature. "I feel ecstasies and inexpressible raptures in blending, so to speak, into the system of beings and in making myself one with the whole of nature. As long as men were my brothers, I made plans of earthly felicity for myself. These plans always being relative to the whole, I could be happy only through public felicity" (Rousseau 1782/2000, 61). For Rousseau the whole always precedes the singular individual, the singular entity, or the singular detail. In society, the common interest precedes the private interest. In nature, the whole ecological system precedes the individuals, whether it be species or organisms. To follow a selfless interest in social and natural science is not only to put aside one's own private or personal interests, but to forget one's own standpoint and try to look at things from the largest relevant totality.

On the other hand, one can also lose oneself in the study of natural phenomena without trying to become wrapped up in the largest possible totality. When one is preoccupied with the study of some fascinating natural phenomenon, it is possible to forget personal interests and to become completely absorbed by the study of the singular phenomenon itself. The disinterested interest is not that of the total ecosystem, nature as a whole, but that of a person who is able to transcend his own selfish interest in the object, and consider the object itself simply as a fellow living being.

This is why the study of botany is more vivid and true to its subject when done by amateurs:

> [A]s soon as we mingle a motive of interest and vanity with it, either in order to obtain a position or to write books, as soon as we learn only in order to instruct, as soon as we look for flowers only in order to become an author or professor, all this sweet charm vanishes. We no longer see in plants anything but the instruments of our passions. We no longer find any genuine pleasure in their study. We no longer want to know, but to show that we know, and even in the forest one is still on the scene of the world, busy to gain admiration" (64).

Professional botanists are no better than any other group of professionals who are eager to compete in the academic world. In their hands the disinterested study of botany withers away like exotic plants in collectors' gardens.

The practice of thinking of plants merely as sources of medicine has also contributed to botany's unsound reputation (60). Botany has suffered much from the fact that the influence of Dioscorides has been stronger in the history of plant studies than that of Theophrastus. Rousseau's introduction to his *Fragments for a Dictionary of Terms of Usage in Botany* begins by stating this in the very first sentences: "The first misfortune of Botany is to have been regarded since its birth as a part of Medicine. As a result people have thought only to find or to attribute virtues to plants and have neglected the knowledge of plants themselves" (Rousseau 1781/2000, 93). For centuries no proper field of research existed that could truly be called botany. The authors of botanical treaties were more interested either in combating diseases, in identifying the herbs found in the classical works of Dioscorides and Pliny, or in discussing words and names, than they were in investigating the structure and internal "economy" of the plants themselves. Only the most useful herbs were therefore explored. It was actually not plants but drugs and remedies the botanists were looking for. The "vegetable chain" was broken down and turned into unconnected links (94ff.).

According to this view, Rousseau points out, Adam must be considered as the first pharmacist, and Eden as the first pharmacist's garden, containing all the drugs of the world, disguised as plants (Rousseau 1782/2000, 60). If this were the whole truth, however, we would be totally indifferent to the members of the vegetable kingdom as long as we were healthy (61).

> These medical ideas are assuredly not very suitable for making the study of botany pleasurable. They whither the diversity of colors in the meadows and the splendor of the flowers, dry up the freshness of the groves, and make greenery and shady spots insipid and disgusting. All these charming and gracious structures barely interest anyone who only wants to grind it all up in a mortar, and no one will go seeking garlands for shepherd lasses amidst purgative herbs. (60)

To treat plants simply as potential drugs is to ignore all those features that make plants truly attractive and fascinating to human beings: their enchanting grace and beauty, their contributions to the atmospheres of places, all their unique and fascinating ways of being alive. But it also distorts botany as a science: the pure philosophical interest in the way the vegetable kingdom is organized, its principles of creation, the basic structure and "economy" of plants, and the curiosity toward all the variety of forms through which life can express itself is crippled and substituted by a petty-minded interest in the medical virtues and capabilities of the most useful plants. Even though one may come to know the drugs quite well this way, one will never know anything about the plants themselves (Rousseau 1781/2000, 94).

It becomes apparent that the disinterested interests in science and in ethics are intimately connected. Botany was inseparable from pharmacy as long as it was ruled by medical interests. It did not become a true science until it released itself from this interest. Similarly, the plants could not be duly recognized in ethics as long as they were looked upon only as potential drugs, balsams, and plasters. The external interest in medical virtues and capabilities excluded or impeded any detached view on the plants. They were always considered as nothing but means to some externally defined goal. Just as science needs to be led more clearly by an ethic of pure curiosity, ethics could use some enlightenment from the curious amateur scientist, who admires all the little wonders of nature that he discovers through his research. Both science and ethics are heading on a wrong track, if they allow medicine or other external interests of utility to decide the rules of the game. In both areas, one must be prepared to treat other life-forms with an attitude of generosity and curiosity similar to the one which Swiss mountain peasants expressed to visiting travellers, or which Julie expressed to the creatures in her Elysium.

Rousseau found these cognated pure interests in science and ethics to be very much in line with his own self-effacing attitude toward his fellow creatures, but at the same time he doubted whether they could be important to more than a few people:

> With respect to that, I feel just the opposite of other men: everything which pertains to feeling my needs saddens and spoils my thoughts, and I have never found true charm in the pleasures of my mind except when concern for my body was completely lost from sight. Thus, even if I were to believe in medicine and even if its remedies were pleasant, I would never find those delights which arise from pure and disinterested contemplation; and my soul could not rise up and glide through nature as long as I felt it holding to the bonds of my body (...) nothing personal, nothing which concerns my body can truly occupy my soul. I never meditate, I never dream more delightfully than when I forget myself. (Rousseau 1782/2000, 61)

The interest in plants must never be confused with external interests, it must remain pure and intellectual, as if we had already been transferred to the Elysian fields.

This does not mean that there cannot be any pleasure in studying plants. On the contrary, there is much pleasure to find. The point is rather that the

greatest pleasure can only be reached when one puts one's own needs and
ambitions aside, even including the longing for happiness itself, and becomes
absorbed in the study of plants as independent creations with a life of their
own.

> Attracted by the cheerful objects which surround me, I consider them, contem-
> plate them, compare them, and eventually learn to classify them; and now I am
> all of a sudden as much a botanist as is necessary for someone who wants to
> study nature only to find continuously new reasons to love it (. . .) In this idle
> occupation there is a charm we feel only in the complete calm of passions, but
> which then alone suffices to make life happy and sweet. (63f.)

Like the mountain peasants who are rewarded more amply when they treat
travellers as friends, the disinterested botanist learns that the best reward can
only be obtained if one does not look at plants as means to satisfy bodily
needs.

Botany is therefore a study most appropriate for the idle life of a recluse,
who has no ambitions for honor and glory, and who has no other intention
with his knowledge than to admire the works of nature or its hidden creator
(Rousseau 1789/1996, 631; 1782/2000, 64). Botany is a study of "pure curiosity, one that has no real utility except what a thinking, sensitive human being
can draw from observing nature and the marvels of the universe" (Rousseau
1781/1979, 106). Any serious defense of biological diversity must take this
as its starting point. There is nothing wrong with utility and bodily pleasures,
but these interests must not get in the way of the pure interest in the phenomena themselves. Like amateur botanists we must develop an attitude of pure
curiosity toward the world of biological organisms. There is a reward to gain
this way, but, like the rewards of love and friendship, it can never be gained
unless we forget about the rewards and become absorbed in the subject of
engagement itself.

Throughout, however, it is not very difficult to find ambiguities in
Rousseau's attitude toward the study of nature. Let me just mention one
of those, which is most directly related to our subject. On the one hand, he is
very much an advocate for scientific studies. In *Émile* it is recommended time
and again to study the laws and creations of nature as directly as possible. One
must encourage the child's curiosity as much as one can, and make the child
familiar with the ways nature works. In the *Reveries* he also points out that it is
necessary to be well acquainted with botanical theory, to understand the laws
of construction in order to experience pleasure in observing them (Rousseau
1782/2000, 64). His own obsession with Linnaeus's sexual system is a consequence of the improved possibilities of observing the constructions of nature

this system has brought him. Rousseau's herbaria were not just condensed diaries of field trips to beautiful landscapes, recalling for his "imagination all those ideas which gratify it most" (68). They were also the basic material, on the basis of which he planned to write botanical treatises on the local flora (Rousseau 1789/1996, 630).

On the other hand, he is quite sceptical about the scientists', especially town-bred scientists,' ambitions. He finds just as much vanity in the world of science as in other branches of culture, and he is continuously critical of the scientists' obsession with artificial systematics, and of their eagerness to isolate the single objects from their natural surroundings in order to put them into botanical gardens, cabinets, and herbaria. It should also be remembered that Rousseau described the idler as the most dangerous man of all in his *Discourse on Science and Art* (1750/1971, 32f.). Science arose from idleness with disastrous consequences: it continuously undermines belief and ridicules ordinary people's commitment to traditional values, it creates a new elite with yet another set of ways to separate people and make them vain and selfish, and it delivers the means to disfigure nature and turn it upside down.

Compared with this he definitely prefers the practical man, in other words, the local peasant's interest in the utility of the plants. There is more honesty and less artificiality in his practical aims than in the professional botanist's interest, especially when it is combined, as it usually is, with vain ambitions for honor and glory. After all, the life of the peasant is "closer to nature" than that of the scientist. The natural way of human beings involves some simple utilitarian interests, and even the least ambitious scientist's detachment introduces yet another artificial barrier between humans and the rest of nature.

This ambiguity can be found, for instance, in the following reflection from *Émile* on the joy of studying the natural phenomena *in situ*, where *in situ* includes the local customs of people who are living "close to nature":

> Is there any one with an interest in agriculture, who does not want to know the special products of the district through which he is passing, and their method of cultivation? Is there any one with a taste for natural history, who can pass a piece of ground without examining it, a rock without breaking off a piece of it, hills without looking for plants, and stones without seeking for fossils? Your town-bred scientists study natural history in cabinets; they have small specimens; they know their names but nothing of their nature. Émile's museum is richer than that of kings; it is the whole world. Everything is in its right place; the Naturalist who is a curator has taken care to arrange it in the fairest fashion; Dauberton could do no better. (Rousseau 1762/1993, 449)

The appropriate setting for plants is not a herbarium or any other artificially ordered collection, but the local environment, which is the creation of the great Naturalist, no matter whether one includes human beings or not.

When Marie Antoinette, in 1775, removed the Parisian botanist Bernard de Jussieu's botanical garden in Versailles, a garden which had been laid out sixteen years earlier in the most systematic way with separate beds for each family of plants, and replaced it with le Hameau, a replica of Sophy's village, it does seem to be very much in line with Rousseau's own considerations, although he would undoubtedly have condemned the artificiality of the whole enterprise. It is the urban scientist's artificial system which is replaced by the very symbol of simple lifestyle. On the other hand, when Julie's husband, Monsieur de Wolmar, ridicules the flower collectors, "those curious little people, those little flower connoisseurs who swoon at the sight of a ranunculus, and bow down before tulips," for their manic obsession with the isolated plants, and argues that flowers should only entertain our eyes in passing, as part of a scenery, and not be expertly anatomized, Rousseau immediately corrects him in a footnote: "The wise Wolmar had not thought about it very carefully. Was he, who knew so well how to observe men, such a poor observer of nature? Was he unaware that if its Author is great in great things, he is very great in small things?" (Rousseau 1761/1997, 395f.).

A similar ambiguity can be found in Rousseau's constant warning against relying too much on books. One must study the natural objects oneself in order to know them well enough. This point can be compared with one put forward in *Émile* by the Savoyard vicar who blames the church and the theology of his day for its indirectness and its reliance on books written by men: "it is men who come and tell me what God has said. I would rather have heard the words of God himself; it would have been easy for him and I would have been secure from fraud" (Rousseau 1762/1993, 312). "How many men between God and me!" he complains (313). Books, botanical no less than theological, only bring second-hand knowledge, and they should always be used with much caution. It is only through direct acquaintance with the natural phenomena that one can become a true student of nature.

But then, on the other hand, it is obvious that the botanical books, first of all those of the Bauhin brothers, of Tournefort, and of Linnaeus, were necessary eye-openers even for Rousseau himself. They made him look for things he would never have found on his own, and they gave him an idea of principles, on which the whole of the vegetable kingdom may be built. The way he organized his own herbaria and the advice he brought forward in his botanical letters were based on lessons from the great botanists. When he later realized that even Linnaeus's system was as artificial, and therefore as empty, as all the previous systems of botany (Rousseau 1789/1996, 631), this cooled his passion for the Swedish naturalist somewhat, but it could not undo the lessons he had learned from him.

In Dr. Thornton's sumptuous and luxurious florilegium, the *Temple of Flora*, published in the beginning of the nineteenth century (Thornton 1807), one can find an illustration showing the celebration of a bust of Linnaeus by a series of allegorical figures, representing various interests in the botanical diversity, which he, Linnaeus, had mapped out for everyone to see. To the left stands Aesculapius with his snake-stick, representing the interest of medicine, and to the right is Ceres, the goddess of agriculture. More to the center of the picture is Flora, representing the sensual beauty of flowers. She is about to put a wreath around Linnaeus's neck, while the god of spring, Zephyr, looking down from the top, strews more flowers on the bust's head. Only the grey head of Linnaeus is present in the picture, but one can easily surmise that the absence of his body and the lack of colors are probably quite suitable for a detached scientist. Cupid represents the sexual system, the most lasting result of Linnaeus's intellectual efforts, and on behalf of the great author of all things Cupid is writing a poem (which the Author allows to be written by the pen of Charlotte Lennox) on the pedestal:

> An animated Nature owns my sway,
> Earth, sea and air, my potent laws obey
> And thou, divine Linnaeus, trac'd my reign
> O'er trees and shrubs, and Flora's beauteous train,
> Proved them obedient to my soft control,
> And gaily breathe an aromatic soul.

Throughout his life, Rousseau was sceptical about doctors, and as we have already seen, he believed that the excessive interest in medical drugs and plasters was to blame for preventing botany from becoming a true science for centuries. But apart from a certain nuisance caused by the presence of Aesculapius, Rousseau probably would have accepted the allegorical picture as a fair representation of the basic interests in the preservation of botanical diversity, and even though he finally was to doubt the naturalness of Linnaeus's system, he would still have approved very much of the celebration itself.

   In Alcinous's and Sophy's parents' garden Ceres seems to have had the most comprehensive role to play, but she always played it in intimate collaboration with Flora. In Julie's garden, on the other hand, neither Ceres nor Aesculapius were ever let in. Flora was an obvious figure here, too, but she was never the main figure. The main part was Cupid's, and he filled it out with utmost skill, never letting the audience know whether the sexual system truly belonged in the science or love department.

When Rousseau himself began to study botany, he was undoubtedly influenced by Flora, but it was the part of Cupid trained in the botany department who soon took the lead. After a while Rousseau almost began to see himself like the bust of Linnaeus. His article on Flowers in the *Dictionary of Botanical Terms* thus opens with the following lines: "If I should let my imagination surrender to the sweet sensations which this word seems to evoke, I would be able to write an article pleasing perhaps to shepherds, but most unacceptable to botanists. So for a moment let us forget bright colours, sweet scents, and elegant shapes, to discover first how to know really well the organism which embraces all these properties" (Rousseau 1781/1979, 134). Like Linnaeus himself, however, Rousseau never put colors, scents, and shapes aside for long. Flora was with him all along, and more so than ever when he was strolling through the countryside searching for new and interesting plants.

Rousseau was always ambiguous and often contradictory in his writings. He was never able to fully reconcile the various images of man as a user of the rest of nature, as a detached observer with a prosaic attitude and scientific ambitions, as an engaged, poetic sensualist who is transported by the beauty and splendor of his fellow creatures, and as a daydreamer lost in reveries of being absorbed in a great unreflective order of nature – to which one must immediately add the image of man as the most valuable creation of nature: the only one who can comprehend and admire at least a few of the traces of the otherwise incomprehensible authorship, and the only one whose freedom allows him to move from innocence to virtue.

If Rousseau had been asked to produce a defense of biological diversity, he would undoubtedly have drawn on all these images, each representing one aspect of human life in need of fulfilment. Plants are both useful and pleasant to peasants like Sophy's parents who were living the most simple and natural life one can imagine. They are interesting objects of study for detached scientists and philosophers like St. Preux, because they each have an independent life. And each of them can be found beautiful or fascinating in their own way, being, like each of us, a creation of the Almighty, as Julie or the Savoyard vicar would have argued.

Rousseau kept insisting that a human being is more valuable than any other organism, due to the gift of freedom, but at the same time he argued that the only way to be worthy of the status as kings of nature, the most appropriate way to use the gift of freedom, is to treat other organisms as fellow creatures, and to underplay the self-centered interests as far as possible. The freedom of human beings should be used in a way which transcends their most self-centered aspirations. Rousseau can thus be considered anthropocentric and nonanthropocentric at the same time: we are only truly human beings if we can avoid being absorbed in our self-centered sentiments, if we can look at other creatures in a detached and yet engaged way. Our freedom allows us to develop a disinterested interest, and this development is the most valuable contribution a human being can make. He was well aware of the contradictions, which could follow from using all the different images of man at one and the same time, but he defended himself by pointing out that "the fault is in nature, not in me" (Rousseau 1789/1996, 174).

Most of the arguments put forward in the modern debate on the value of biological diversity can already be found in the writings of Rousseau: utilitarian value, aesthetic value, scientific value, amenity value, transformative value, existence value. I hope I have shown that he is still able to make an interesting contribution to the debate. At the same time, however, it is necessary to avoid the most fundamental contradiction in his theory: we cannot continue to use human-free nature as a normative standard, while at the same time insisting on the freedom of human beings as the crown of creation. The most promising way to avoid the contradiction is not to become misanthropic and deny that humans have anything valuable to contribute, but rather to insist that human beings are part of nature, that a human creation can be just as valuable as any other natural phenomenon, and that we can never fully move beyond the point of view we have inherited as natural creatures living at a particular place in time and space. Human culture is yet another part of the biological diversity, and if people like Rousseau are right, it is likely to flourish much better, and be qualitatively better off, if it is able to develop a disinterested interest in species of organisms that are different from us . . . even though we cannot avoid eating carrots and beans, or even a medium-done steak now and then.

BIBLIOGRAPHY

Arler, F. 1997. *Ind i naturen eller ud af naturen? Om Rousseau og den naturlige opdragelse.* [Into nature or out of nature? On Rousseau and natural education]. Odense: Humanities Research Centre "Man and Nature."
Homer 1994. *The Odyssey.* Trans. S. Butler. London: Barnes and Noble.
Morton, A. G. 1981. *History of Botanical Science.* London: Academic Press.
Rousseau, J. J. 1750/1971. *Discours sur les Sciences et les Arts.* In *Schriften zur Kulturkritik*, K. Weigand, ed. Hamburg: Felix Meiner.

1762/1993. *Émile.* Trans. B. Foxley. London: Everyman/J. M. Dent.

1761/1997. *Julie or the New Heloïse.* P. Stewart & J.Vaché, eds. Hanover and London: University Press of New England.

1781/1979. *Botanical Letters* (written 1771–73) and Notes towards a Dictionary of Botanical Terms. In *J.-J. Rousseau: Botany. A Study of Pure Curiosity,* R. McMullen, ed. Illustrated by P. J. Redouté. London: Michael Joseph.

1782/2000. *The Reveries of the Solitary Walker* (written 1776–78). Trans. C. E. Butterworth. In *The Collected Writings of Rousseau Vol. 8.* C. Kelly, ed. Hanover and London: University Press of New England.

1789/1996. *The Confessions* (written 1766–70). (Translator anonymous.) Ware, Hertfordshire: Wordsworth.

1781/2000. *Fragments for a Dictionary of Terms of Usage in Botany.* Trans. A. Cook. In *The Collected Writings of Rousseau Vol. 8.* C. Kelly, ed. Hanover and London: University Press of New England.

Thornton, J. R. 1799–1807. *Temple of Flora.* London: J. R. Thornton.

7

There Is Biodiversity and Biodiversity

Implications for Environmental Philosophy

KEEKOK LEE

This chapter argues that biodiversity is what may be called a secondary, rather than a primary, ontological characteristic – a distinction that is borne out and required to be made by the increasing ability on the part of humankind, through the technology of genetic engineering, in principle, to create new organisms with genetic material belonging to a very different species inserted into their genome.[1] If biodiversity per se is either only of instrumental value or is a mere secondary characteristic, then such human ingenuity in manufacturing its own biodiversity would not be troublesome. But if nature's biodiversity, as an expression of nature's own creativity, is a primary, ontological characteristic, then the displacement or transcendence of naturally occurring biodiversity by humanly fabricated biodiversity would be philosophically troublesome. However, this should not be misinterpreted to imply that no powerful instrumental reasons for caution about humanly fabricated biodiversity exist. Clearly, they do.[2] However, it is not the objective of this contribution to demonstrate these more familiar worries and concerns. Instead, the focus is on the less commonly articulated ontological objection to the loss of naturally occurring biodiversity. But the point made presupposes the related distinctions between human and nonhuman, on the one hand, as well as between nature and culture on the other.

7.1 BIODIVERSITY: ITS COMPLEXITY

"Biodiversity" as a term refers to a complex set of phenomena, as biodiversity takes more than one form. In its popular sense, it is about species diversity, that is, the number and existence of different species of plants and animals today, estimated to vary from 3 to 10 million at the conservative end, to between 10 and 100 million at the more speculative end. However, Plants and Animals (which are multicellular eukaryotic organisms whose DNA is contained in a well-defined cell nucleus with a protein coat) are only two out of five kingdoms; the others are Monera (the prokaryotic organisms like bacteria), Protista (the unicellular eukaryotic organisms like protozoa whose genetic material is not contained within a well-defined nucleus), and Fungi (the multicellular eukaryotic organisms). As biodiversity covers these as well, the number of extant species would rise dramatically. However, knowledge of these species is even more sketchy than of plant and animal ones. Work on the bacteria has only just begun. Furthermore, what counts as a species itself is not straightforward. Biologists tend to operate with the biological species concept which applies best in the case of sexually reproducing plants and animals, but is not applicable to asexually reproducing ones, or to bacteria in the Monera kingdom.[3]

Biologists are also interested in genetic diversity which occurs at different levels of biological organization. Each species varies in the amount of genetic material it contains – the bacteria have about 1,000 genes, some fungi 10,000, many flowering plants 400,000 or more. Furthermore, each species is composed of numerous individual organisms whose genetic components are not identical (except in cases of monozygotic twins and parthenogenesis of asexual organisms). Other biologists are concerned with different populations of a species and the genetic variations that exist both within as well as between the populations. But in any case, the loss of genetic information, whether in the case of one individual organism, the loss of a breeding population within a species, or the extinction of a species, is an irreplaceable loss.

However, genetic diversity, though clearly fundamental, should not be identified simplistically with biodiversity as the latter goes beyond the former. A species, a population of individual organisms, or an individual organism is not merely a stock of genes. Each ultimately is a member of an ecosystem with which it dynamically interacts – the genetic information it presently carries is itself the result of past adaptations to environmental forces and represents possibilities for future adaptations. From the ecological as well as evolutionary points of view, a species or a similar population of individual organisms in two different ecosystems represents two different ecological and evolutionary potentials. Ecosystem diversity is, therefore, a crucial aspect or form of biodiversity.

However, before proceeding further, let me enter a caveat – this chapter, in setting out and defending its main thesis, may invoke only certain forms of biodiversity in its arguments; but this narrow focus should not be interpreted to mean that those arguments do not also apply to the other forms which it fails, for lack of space, to look at in detail.

7.2 TECHNOLOGY AND HUMAN-MADE BIODIVERSITY

The biodiversity referred to in the previous section is what may be called naturally occurring biodiversity.[4] However, this chapter wishes to argue that from the philosophical/ontological point of view, it should be distinguished from another kind, namely, human-made biodiversity. To make this clear, one must next say something about biotechnology, the current, much acclaimed technology which is used to transform the former to become the latter.

Such transformation has, of course, a very long history, beginning with the first domestication of plants and animals.[5] However, from that dim and distant past, spanning several millennia to the 1930s, the technology relied on by societies and cultures in various parts of the world is what may be called craft-based technology. (However, this is not meant to imply that the results achieved were not thoroughly impressive and sophisticated. To grasp the enormity of such achievements, just consider and compare the wild ancestors of wheat (domesticated in the Fertile Crescent of the Old World) and maize or corn (domesticated in the New World) with their respective domesticated counterparts). Its methodology of artificial selective breeding consists mainly of trial and error. Insofar as there was theoretical speculation about the phenomenon, it was at best fragmentary, not really systematic; nor did it provide any satisfactory explanation to account for the practical success achieved.

However, the history of domestication took a radical step forward with the so-called rediscovery of Mendelian genetics at the beginning of the twentieth century which, for the first time, established the modern science of genetics itself. Mendel's law of segregation shows that in the transmission of characteristics (at least in sexually reproducing organisms), offspring inherit characteristics from both parents that are passed down further in the line in certain clear ratios, even though they may not be manifest in the individual organism itself.[6] These characteristics are transmitted via genes that are located in chromosomes which vary in number depending on the species to which the individual organism belongs. The gene/chromosome theory provides the long sought after explanation to account for the phenomenon of hereditary transmission, discrediting in the process competitors like the theory of blending inheritance or that of Lamarck, which postulated the transmission of characteristics acquired in the lifetime of the individual organism.

The gene/chromosome explanatory framework was immediately appreciated and seized upon for its practical/technological potential by farmers and breeders (initially the American ones), who realized that it could be made to yield far more precise and accurate results than the traditional craft-based breeding methods based on trial and error. For the first time in the history of domestication, a fundamental discovery in basic theoretical science could be used to induce or generate a superior technology for the manipulation and control of genetic material. It took roughly thirty years for the double-cross hybridization technology to be perfected which radically transformed agriculture (which up to then had relied on so-called landraces) and husbandry. Individual plants and animals came to be collected around the world for their genes by what today we call First World economies, and brought back to centers of research in such nations in order to cross-breed and develop new varieties of plants and animals bearing certain desired characteristics. Modern agriculture and husbandry were born; the most talked about version of which in the former is the so-called mid-twentieth-century "Green Revolution" of high-yield rice and wheat, among other crops.

For the first time, experimentation in plant (and animal) breeding was no longer in the hands of farmers. Instead, it passed to research laboratories, either in universities funded by the government or in private seed companies, that is, to scientific experts who understand the new science behind the new technology.

Although the gene/chromosome theory focuses on particular trait(s) of the individual organism (belonging to one variety) deemed by the breeder to be desirable and which is to be bred into another plant/animal (belonging to another variety), the actual breeding nevertheless takes place between two individual organisms – the actual mating could take place naturally (but in an environment controlled by the breeder) or with the help of the breeder, such as in pollination by hand, or more recently, in vitro fertilization in the case of animals. However, today, this relatively reductionistic technology has been, if not displaced or totally transcended, augmented by yet another, even more reductionistic and powerful technology commonly called biotechnology. (Note that the term "biotechnology" as used and understood today is, in many popular minds, more or less synonymous with that technology referred to also as "genetic or DNA engineering" which is induced by the basic theoretical science of DNA or molecular genetics. However, it goes beyond mere genetic engineering as it includes other biogenetic technologies like in vitro fertilization, embryo selection, cloning [of animals, not merely plants] which are induced by theoretical understanding given by such a basic science as cell biology. The term has also been used extremely broadly to include all historical forms of human manipulation of organisms including traditional craft-based technologies like brewing/fermentation, domestication of plants and animals via artificial selective breeding, double-cross hybridization [a breeding technology informed by Mendelian genetics and chromosome theory] as well as DNA engineering. This comprehensive definition is not recommended as it causes confusion.)

Genetic engineering (as one key component of biotechnology) may be said to be a more powerful technology than double-cross hybridization precisely because DNA genetics is a more basic science than Mendelian genetics. In its explanatory schema, DNA genetics goes beyond genes/chromosomes to the structure of the DNA molecule which constitutes the genes themselves. Crick and Watson in identifying that structure also made it possible for humankind for the first time to manipulate genetic material at the molecular level.

Of course, in order to put genetic engineering in place, an immense amount of additional work, both experimental and theoretical, had to be done first – such as the development of restriction and other enzymes, of vectors and "vehicles" to deliver the DNA, of quick and efficient techniques to create gene libraries, including the human genome. But within about twenty years of the discovery of the double helix, the new technology has been established. It differs from the Mendelian-induced technology in at least two radical ways: (a) the whole individual organism can now be superseded for the purpose of producing organisms with certain so-called desirable characteristics. All that is required is the insertion (in those cases where single genes are involved) of the appropriate sequence of DNA from one organism into the genetic material of another organism; (b) the donor and receiving organisms need not be varieties of the same species, but may belong to two distinct and different species, be these plants or animals – a strand of DNA of one species of plant or animal may be inserted into the DNA of a plant or animal of a separate species, just as the DNA of one plant species may be spliced into the DNA of an animal species, and vice versa. In other words, genetic material which does not meet and interact in naturally occurring situations is now brought together to produce what is commonly called "transgenic organisms." The truly radical nature of this breakthrough is that by manipulating and control-ling genetic material at the molecular level, it is now possible to cross the species, indeed, even the kingdom, barrier. Such a technology entails in principle, if not yet quite in practice, the systematic supersession of natural evolution itself. (However, this remark does not apply to humans. Human reproduction is part of cultural, rather than natural, evolution. In any case, at the moment, many of the techniques and protocols, successfully developed, for other animals have not yet been perfected for humans; such experimen-tal imperfection may persist unless what are perceived, now, to be ethical or religious constraints to extending the techniques to humans no longer obtain.)

7.3 BIODIVERSITY AND INSTRUMENTAL VALUES

Obviously, biodiversity may be valued instrumentally. However, to say this is not to say that it is, or can be, valued solely in instrumental terms. To hold that it can only be of instrumental value constitutes the philosophy of instrumentalism within an anthropocentric framework – if humankind alone is intrinsically valuable, it follows that (nonhuman) nature, including biodiversity, can only have instrumental value for humans. The usual arguments cited in favor of saving biodiversity from this perspective are:

a. resource value (resource conservation) – the silo argument;
b. amenity value (resource preservation) – providing (i) psychological, spiritual, or religious sustenance (the cathedral argument); (ii) recreational opportunities (the gymnasium argument); (iii) aesthetic pleasure (the art gallery argument);
c. biospheric value (resource preservation) – as provider of "public services and goods" by maintaining the great natural cycles such as the hydrological, and by acting as a sink to absorb the waste and pollution produced by human production and consumption, without which human life is not possible (the life-support argument).

In other words, the loss of biodiversity is worrysome because ultimately human material/economic well-being could be adversely affected; human health may be endangered if biota with potential medicinal properties were to become extinct; human psychical well-being may also be undermined. Humankind should therefore be prudent in harvesting, using, and managing biota to ensure all-around flourishing of the present as well as of future generations. This constitutes the philosophy of conservationism (which covers both resource conservation and resource preservation, according to the terminology used in this chapter).

The most important causes of biodiversity loss today are anthropogenic – the pressure from human population growth, habitat fragmentation and destruction, hunting and harvesting, pollution as well as atmospheric and climatic change, and the introduction of exotics.[7] Before humans appeared, the (nonanthropogenic) rate of species extinction – which is only one measure of biodiversity loss – is said to be one organism per one thousand years. But since 1600, in the space of four hundred years, over one thousand species have become extinct. Since 1960, one thousand are lost per year. The rate at the end of the twentieth century was said to be one species per hour, but even this has been regarded by some experts, like E. O. Wilson, to be on the low side, who cite the figure of three per hour. Obviously, the predicted rate depends on, among other factors, the estimated number of species believed to be extant. But whatever the estimated figure, the community of biologists agrees that biodiversity loss in general has accelerated at an alarming rate, especially of late, caused in the main by habitat fragmentation and destruction, induced by pressure from human population and its attendant demand, particularly in the industrialized economies, for ever-increasing use and consumption of natural resources.

7.4 CONSERVATION BIOLOGY AND ITS 'DIRTY SECRET'

Although instrumentalism and conservationism (as previously defined) are the dominant, and therefore, the intellectually respectable, attitudes toward biodiversity, there remains a stubbornly different though usually suppressed response which, for instance, exists among many conservation biologists themselves who nevertheless admit to it only when pressed (see Takacs 1996). They justify their interest in and secure funding for their research by relying on the rhetoric of instrumentalism. Yet privately, they are not convinced that species and their ecosystems only have instrumental value for us. For both pragmatic and philosophical reasons, they keep to themselves a kind of "dirty secret" which should not and could not, under normal circumstances, be mentioned in public.

But if the philosophy of instrumentalism truly exhausts the debate, then biotechnology, which is expected to take the industrialized economies into the twenty-first century, could be embraced without hesitation. The silo justification could, for instance, be effectively met by storing DNA sequences in laboratories, instead of storing actual whole seeds (in the case of plants) in cryotoria (as seed banks) which would have a relatively shorter shelf life than storage as DNA under optimal conditions. Furthermore, DNA sequences can be more readily replicated than the seeds themselves. What must be preserved for human flourishing is DNA sequences, not the individual seed or egg/sperm, the individual organism, the populations of organisms which make up the species, nor the particular ecosystems of which these individuals and populations are members.

Again, take the life-support justification. Tropical rainforests (where biodiversity is densest) play the role of absorbing carbon dioxide emitted by our productive/consumptive activities. If they were systematically cut down, such a biospheric function would be upset, and further environmental problems would also ensue, like soil erosion. But research is being carried out to fabricate trees, genetically engineered to absorb even greater amounts of carbon dioxide than the naturally occurring ones in the Amazon today. If and when such research comes to fruition, no loss of value would be incurred in cutting down the forests themselves provided they are replaced by the more efficient substitutes. Today, the tropical rainforests still persist to an extent in spite of the deforestation that has already occurred. One reason for its survival to date is that no sustainable agriculture is possible on its poor soil. But biotechnologists are already working to genetically engineer crops that can grow sustainably on such soil. When that comes to pass, the pressure to develop the forests (or indeed "to green" deserts with analogous suitably genetically engineered plants) would be overwhelming.[8] In principle, if not yet already in practice, it is now possible to fabricate biodiversity if and when we see fit. (Looking even further into the future, for those focusing on space colonization, Mars could be terraformed. Although the planet may have no life and, probably, had no life [in spite of speculation to the contrary], when an atmosphere like Earth's has been manufactured, biotechnology could genetically engineer special organisms to fit such an atmosphere and speed the process of transforming Mars to become an *ersatz* Earth.)

7.5 THE ONTOLOGICAL DIMENSION

But what if the philosophy of instrumentalism and conservationism does not exhaust the debate about the value of biodiversity? In turn, what threat would biotechnology and its enfolding potential pose to biodiversity itself? The remaining sections of this chapter will address such matters and attempt, here, a mere sketch of a possible defense for biodiversity other than in instrumental terms (for a fuller defense, see Lee 1999).

In the discourse of environmental ethics, the concept of instrumental value is said to be the antonym of intrinsic value. (Indeed, some theorists go further to say that the notion of instrumental value entails that of intrinsic value; otherwise, it would lead to an infinite regress. Traditionally, as we know, this regress is stopped by postulating that humans are uniquely intrinsically valuable because they possess souls, consciousness, self-consciousness, language, and so on.) But this author prefers to invoke the notion of independent value instead (for reasons which will soon become obvious). Minimally it is to be understood as follows: an entity (or process) is morally considerable and has value if that entity has come into existence, continues to exist, and then goes out of existence – in principle – entirely independent of direct human intention, design, manipulation, or control. The operative phrase here is "direct intention." It is undoubtedly true that human culture itself has provided suitable ecological niches for certain species and their evolutionary changes. For instance, sewerage systems in large modern cities would, and could, have encouraged different populations of rats to thrive and flourish, each with very different evolutionary potential. However, such species are not artefactual within the meaning of the definition given here. Sometimes, in initiating human activities, no one could have foreseen certain consequences; these become apparent perhaps only centuries later. As they were not even foreseen or foreseeable, they could not be said to be intended. At other times, one could perhaps foresee such consequences, but as the direct goal is not to bring them about, they can at best be said to be only indirectly intended. Back to the sewerage and rats example. The explicit goal of designing and laying down a modern sewerage system is to dispose of the detritus of crowded living, not to enable rats to proliferate and to evolve under such conditions. Nor are those species artefactual which are only viable in the various ecological niches provided by the human body. In our skin and in our guts thrive numerous species of micro-organisms, which, in the absence of humans, would not have evolved and/or survived. But these, too, are not artefactual, as their existence has nothing to do with human intention, direct or indirect.

In other words, for an entity to qualify as possessing independent value, it would or could have existed on Earth before the appearance of *Homo sapiens* in evolutionary history, and would or could exist if humankind were to become extinct.[9] With this understanding, the possession of consciousness, self-consciousness, or language which is said to invest intrinsic value in humans who are the sole bearers of such characteristics (in anthropocentric thought) is beside the point – even rocks and plants which lack consciousness qualify to be the loci of independent value.

Focusing on this value makes it easier to see that instrumentalism is flawed. Entities have not come into existence in order to be of use only to humans. Their comings and goings are entirely independent of human intentions or manipulation. When humans finally appeared on the evolutionary scene, they, like other species, obviously found and have continued to find some of these entities and items to be of use to them, and without which they would and could not continue to survive, never mind flourish. In this sense, of course, certain plants and animals, the air, water, the soil, all have instrumental value for humans. But to admit that they have instrumental value for humans is not the same as saying that they have no other value except to be of use to humankind. The former amounts to uttering a truism. The latter is an articulation of the philosophy of instrumentalism itself, which seems to follow as an entailment of the thesis that humans are the sole locus of intrinsic value. Or it is an entailment of the claim – what may be called external teleology, first enunciated by Aristotle in *The Politics* – that nonhuman entities have come into existence solely for the purpose of serving human goals and purposes. However, the first entailment may be challenged, as nonhuman entities may be the loci of independent value.[10] The second entailment does not follow simply because Darwinism and neo-Darwinism have long exposed the thesis of external teleology to be empirically false as well as metaphysically suspect. The mechanism of natural selection working upon genetic variations in individual organisms constituting populations of organisms in their interactions with their ecosystems primarily explains the course of evolution. Neither divine nor human intentions infuse or guide it. As already observed, it is a matter of mere contingency that humans find certain entities, but not others, useful to their survival and flourishing.

The concept of independent value introduces the ontological dimension into the discourse of environmental philosophy.[11] It implies that ontologically speaking, two very different types of entities exist on Earth. On the one hand are those that have come into existence, continue to exist, and go out of existence entirely independent of humans, their intentions, and manipulation and, on the other, those entities that have come into existence, continue to exist, and go out of existence precisely as the result of human intentions and manipulation. The former category covers naturally occurring entities, where the term "nature" is understood in its sense as the ontological foil to the latter category which refers to human artefacts.[12] Human artefacts may be briefly defined as the material embodiment of human intentionality; as such, they are technological products, whether the technology involved is craft-based or science-induced.

The ontological difference between naturally occurring entities and artefacts has already been intimated but may be brought out more explicitly by reminding the reader of two things: (a) in a world before the appearance of humans or after the disappearance of humans, the first kind of entities had existed and would continue to exist; (b) only in a world with humans would the second kind exist. The second remark, however, is not meant purely as an empirical but also a conceptual one – in a world without humans, the notion of (human) artefacts is unintelligible. We humans in creating the artefacts – be these utilitarian objects like cups and shoes in daily use, or artistic ones like paintings, or religious/spiritual ones like temples and cathedrals – endow them with meaning. While we throw away the functional utilitarian ones after a while, we tend to cherish the others, carefully maintaining and repairing them against the ravages of wear and tear through time.

These points could be brought home more vividly through a thought experiment, based on an adaptation of the Last Person Argument. Imagine yourself to be the last person on Earth. After your demise, no other human or being with a consciousness equivalent to that of humans would ever arise. It is feasible for you to set up a destruct device programmed to go off as soon as you had expired, blowing up both biotic and abiotic nature (the latter in the form of geological formations like extant mountains, rivers, and lakes) into smithereens. Are you morally permitted to do so? The answer is obviously "yes" if the philosophy of instrumentalism is unswervingly correct; but it may be "no" if both biotic and abiotic nature have a value that is independent of us humans. However, this insight should not be distorted to mean that nature is static, that in not destroying it after the last person had died, nature – biotic and abiotic – would not change. Of course it would, but the direction and pace of change is not dictated by humans, it is independent of us. In destroying it, the last person would be dictating its fate, interfering with its own trajectory.[13] It should be allowed to carry on autonomously when humankind is gone. (Note, however, that this sense of autonomy carries no Kantian baggage, and may, in many contexts, be synonymous with the word "autopoiesis.")

Now, consider another scenario. Suppose it is possible for the last person to arrange for the destruction of all (human) artefacts after his or her demise. (However, a qualification to this will be made in a later section.) Is it morally permissible to do so? The answer must be "yes." Artefacts are fabricated by humans, embodying our intentions, designs, and purposes. In the absence of human consciousness, there would be no Taj Mahal, no *Mona Lisa*, no shoes nor saucepans, only piles of marble/stone/brick/wood, bits of fibers and pigments, of old skin, or collections of metals/minerals. Ultimately, they would all turn back into the elements of carbon, nitrogen, or whatever they were basically made of. It would make no sense to protect them from either instant destruction or eventual decay in the absence of humans. Apart from the instrumental value such artefacts have for us, be these mundane or spiritually and aesthetically uplifting, they can have no other value. Nor does it make sense to say that they have. When we are gone, the ants, the spiders, the micro-organisms crawling over the canvas would not be able to appreciate them as our creative products; to some of them, the canvas might be a source of nourishment, to others, it might provide an ecological niche, which would not be available if the canvas had not been made and left by us.

7.6 UNIQUENESS OF HUMAN CONSCIOUSNESS

Human consciousness differs radically from the types of consciousness found in nonhuman others which share Earth with us. This is as true of chimpanzee consciousness as it is of the more lowly vertebrates like fish (whose type of consciousness probably only permits them to feel pain but not to form the desire to end it) (see Varner 1998, 26–54). Those who are overly anxious to distance themselves from the charge of anthropocentric arrogance keep reminding us that humans and chimpanzees share 98.4 percent of DNA, and that our ape cousins are really no different from us and are truly human.[14] However, this is a mistake, born of genetic determinism or reductionism, among others. What makes human consciousness so different from that of the chimpanzee variety lies precisely in that small difference in genetic material. This difference manifests itself in a new form of mental activity, not seen before so far in evolutionary history, namely, the ability on the part of humans to recognize not only our own creativity (as expressed in the artefacts we fabricate) but also (nonhuman) nature's creativity as well. Such a consciousness is also able to grasp that while our artefacts have value for us (as an expression of our intentionality and purposes), there may also be (independent) value in nonhuman others that has nothing to do with our intentions or manipulation. The creativity that the chimpanzee is capable of expressing constitutes a value, not because the chimpanzee is very close to us in its genetic constitution, but because it has evolved and come into existence entirely independent of us. Similarly, the creativity of plants (which have no consciousness whatever) in ensuring their own survival, reproduction, and flourishing has nothing to do with us; however, it is not obviously unintelligible for humans to recognize and appreciate such creativity. The chimpanzee, on the other hand, is blissfully ignorant that it itself embodies or enacts such creativity. It cannot recognize or articulate such a value either in itself or in other species. Nor can it appreciate the destruction of such a value in others as a moral evil, as chimpanzees lack the intellectual capability to be moral agents.[15] But humans are moral agents, uniquely so. (Of course, not all humans are moral agents; but all moral agents are humans. However, it is the latter, not the former, proposition [both equally true] which is germane to the argument here.) While we are capable of appreciating and articulating the instrumental value which nonhuman others may have for us, we are, at the same time, also capable of appreciating and articulating that independent value which is embodied and exhibited in nonhuman others. For this reason, it is right and proper that we should agonize about our destruction of existing naturally occurring biodiversity, about the possible and potential substitution of such naturally occurring biodiversity by our latest technological means, namely, biotechnology, which has been induced and spawned by our basic theoretical understanding through molecular DNA genetics and cell biology.

There is a need to distinguish between what may be called primary and secondary characteristics which an entity (or process) bears or exemplifies.[16] The former is said in this context of discussion to be primary because it is the bearer of an ontological dimension. For instance, creativity, a value referred to in the last section, is capable of being exemplified in two very different kinds of context. So can other values like complexity or ingenuity. Creativity as manifested by us humans, in our artefacts, is of a different ontological order from the creativity as manifested in naturally occurring nonhuman entities and processes, although it appears superficially to be one and the same value. The former has come into existence, continues to exist as the material embodiment of human intentionality, and will go out of existence at the behest of human goals and purposes; the latter has come into existence, continues to exist, and will go out of existence, independent of human design, manipulation, or control. In other words, it is misleading to speak of creativity or complexity *tout court*. Instead, one should be sensitive to the following questions: "whose creativity?" or "what is the source of the complexity?" These questions must be understood and grasped before one can adequately and fully assess the value of creativity or complexity itself.

From the ontological perspective, what is crucial is the context in which the value of creativity, complexity, or ingenuity occurs. In other words, human creativity, complexity, or ingenuity is not substitutable for naturally occurring creativity, complexity, or ingenuity. In the case of biodiversity, suppose it is possible in practice, not merely in principle, one day, for biotechnology (in synergistic conjunction with other technologies, like microcomputer technology and/or molecular nanotechnology) to fabricate equally complex, ingenious organisms to replace the naturally occurring biodiversity that exists today. However, such an eventuality does not imply that there is no loss of value. The implication only holds if the focus is entirely on the secondary values of creativity, complexity, or ingenuity, but fails to hold when the focus is shifted to the ontological dimension. Nature, as "ontological other," would have been displaced or destroyed should we systematically use radically powerful technologies, like biotechnology, to transform the natural to become the artefactual.

When we admire the creativity, complexity, or ingenuity in the biodiversity found, say, in the tropical rainforest today, we are admiring nature's own creativity, complexity, or ingenuity. But when we admire (and there is a lot to admire) the creativity, complexity, or ingenuity in the cultivated maize as originally domesticated by the American Indians, which bears little or no resemblance to its wild ancestor, we are, in reality, admiring our own creativity and ingenuity. We, humankind, have cleverly and successfully transformed the plant in the wild to become an entity radically different from what it was, in order to embody our intentions, desires, and purposes. In other words, all cultivars and all domesticated animals are biotic artefacts. Whether we use the traditional craft-based techniques of breeding based on a good deal of trial and error, the more sophisticated Mendelian science-induced technology of double-cross hybridization, or the most recent, even more sophisticated DNA science-generated genetic engineering, what we have done is to simply ensure, with increasing precision, that certain characteristics the plant or animal bears which we deem to be undesirable (from the standpoint of our purposes and desires) are eliminated from its genotype, while those characteristics we deem to be desirable are bred into or inserted into its genotype.

From this point of view, the domesticated organism is, therefore, every bit as much of an artefact as the statue of Queen Victoria standing in the square of an English provincial city.[17] The fact that such an organism breathes, metabolizes, and reproduces is immaterial, for such mechanisms have been hijacked by us humans to do a job which we want them to perform, namely, to breathe, eat, digest, and reproduce itself, on our behalf, according to our specifications, and not its own. In Aristotle's language, one would say that it no longer enacts its own *telos* – the sheep which has been genetically engineered to produce in its milk, not its own hormone, but a human one, could not be said, in one crucial sense, to be unfolding its natural *telos*. Admittedly, it is true that humans are not responsible for the fact that the transgenic organism continues to breathe, eat, digest, defecate. At a superficial level, it seems to be autonomous, discharging its natural *telos*; but at a deeper level, it has lost its autonomy as humans have captured that *telos* to do what we humans want it to do, to produce a human hormone in its milk which, left to its own biological devices, would never have come about. It is human subversion at such an extremely deep level of manipulation which renders it so thoroughly an artefact.

In this context, one then needs to distinguish between yet another two theses of teleology apart from the thesis of external teleology already identified – intrinsic/immanent teleology, on the one hand, and extrinsic/imposed teleology, on the other. The naturally occurring organism embodies the former but the domesticated organism the latter. Humankind has substituted extrinsic/imposed teleology (such that organisms do or exhibit characteristics humans deem to be desirable) for intrinsic/immanent teleology (under which organisms do what they do and/or exhibit characteristics as these have naturally evolved, regardless and independent of human intention and volition).

Another way of appreciating the points just made is to define the notion of artefact somewhat differently from the account at the beginning of the chapter. One leans once more on Aristotle, whose explanatory schema of phenomena consists of the four causes: material, efficient, formal, and final.[18] An abiotic or exbiotic artefact, like a statue, is to be explained or accounted for by reference to its material cause (the early technologists fabricated bronze an alloy which does not occur naturally but is made out of copper and tin), efficient cause (the artist made it), formal cause (the artist created it bearing in his head the image of the Queen Empress or in accordance with a sketch of the subject), final cause (the nation's intention to celebrate Victoria's Golden Jubilee). In principle, each of these four causes can be assigned to four different external (human) agencies. In contrast, Aristotle said that in the case of the (naturally occurring) organism, the four causes may only be distinguished in principle – intellectually and theoretically – for such an organism is itself its own material, efficient, formal, and final causes. However, in the case of the domesticated organism and, even more so, of the genetically engineered type, the four causes, like those of the statue, can be assigned to four different external (human) agencies. Its final cause is designed by us (that the sheep should manufacture a human hormone); its formal cause is in accordance with a particular human blueprint; its efficient cause is humans who have brought it into existence with the characteristic(s) it possesses; its material cause is assembled by us, humans, who ensure, either through selective (artificial) breeding or by insertion of DNA sequences, the genotype it possesses.

One must enter a caveat here, as promised. In the Last Person thought experiment mentioned earlier, this contribution has argued that it is morally permissible to destroy all (human) artefacts after the expiration of the last human. But in the case of biotic artefacts, one could argue for an exception to be made and that they be spared. In the absence of humans to sustain them, many of them would, undoubtedly, perish or be unable to reproduce successfully without human assistance (like camels or turkeys) even if they could survive. But conceivably some may survive and reproduce themselves, like feral pigs or horses. Over long stretches of geological time, the degree of their artefacticity would attenuate while still bearing, in their history, their origins as human artefacts. Without humans, extrinsic/imposed teleology would recede and intrinsic/immanent teleology might have a chance to re-assert itself in some cases.

7.8 ANTHROPOGENIC AND NONANTHROPOGENIC

In light of the preceding discussion, it is also necessary to explore a little further the philosophical implications of the distinction between anthropogenic and nonanthropogenic in the context of species extinction and biodiversity loss in general. Some writers have rightly pointed out that in Earth's history there have been at least five major extinctions, well before the appearance of *Homo sapiens*. Each time Earth recovered its biodiversity, although recovery took millions of years in each case. However, from this undeniable fact, they wrongly inferred that the human extinction of species is of the same order as such natural episodes, thereby attaching no philosophical significance to the distinction between anthropogenic and nonanthropogenic biodiversity loss.[19] To them extinction is extinction *tout court*. But biodiversity loss is a secondary disvalue. It is important to focus once again on the ontological dimension – it matters whether the loss is caused by us, humans, or not. The two kinds of extinction are not of the same ontological order. Nonanthropogenic loss in biodiversity has no moral significance whatsover. But the different ontological order of anthropogenic biodiversity loss carries moral resonance. We have already seen, in an earlier section, that human consciousness is unique precisely because it can agonize about, and appreciate, that nature, as "ontological other," has independent value, that the destruction of such a value may be said to constitute moral evil, as such destruction is simply a systematic attempt on the part of humankind to destroy or impose its will on other nonhuman forms of existence. Constraints against such wholesale destruction or systematic transformation of the natural to become the artefactual within the philosophy of instrumentalism are entirely prudent. But should biotechnology (and other technologies of the rising future) permit us to overcome them, then expediency would bid humankind to continue going down this imperialistic path. Breaking free from instrumentalism might provide ontological *cum* moral constraints which should make us pause, and perhaps, even pull back from a headlong rush to transform and humanize nature via our technologies which, since the 1840s, have been generated by the theoretical understanding of nature given by modern science.

7.9 DIFFERENT SENSES OF 'NATURE'

It is necessary to prevent a certain misunderstanding arising from the arguments set out in this chapter. The arguments do not deny that humans, in one sense of the term "natural," are natural beings – this is the sense whose antonym is "supernatural." Humans, not being gods or angels, are not super-natural but natural beings. Like other organic natural beings, they are part of Earth's biotic evolution. This sense may be called "natural$_c$" and include all things with spatial/temporal coordinates, whether these be humans, beavers, the Hoover Dam, or the beaver's dam. However, the term "natural" has other meanings which must be distinguished and grasped if the issue about either the loss of naturally occurring biodiversity through human activity or the increase of humanly fabricated biodiversity as substitute is to be properly understood. But for the purpose of this discussion, only one other of these senses may be said to be immediately germane, namely, "natural$_{fa}$," which is the ontological foil to the concept of the artefactual. In other words, what is natural in this sense is what is not artefactual.

This sense in turn relies on the related distinctions, human/nonhuman on the one hand, and nature/culture on the other. These distinctions are neces-sary given that humans possess a unique consciousness, as already observed in section 7.6, which sets them apart from nonhumans. Among other things, it permits them to achieve an extremely high intellectual level of development as manifested in today's science of molecular biology (including molecular genetics) and the technologies derived from it called biotechnology. Biotech-nology enables humans to create transgenic organisms which, *ex hypothesi*, are not naturally evolved beings (in the sense of "nature$_{fa}$," though natural in the sense of "nature$_c$"). As such, humankind, though part of nature$_c$, never-theless, falls outside nature$_{fa}$; what it has accomplished through its technolo-gies are part of human culture, not nature$_{fa}$. Transgenic organisms are biotic artefacts as they are the material embodiment of explicit and direct human intentions; in contrast, hybrids between varieties (within a species) in the wild are part of nature$_{fa}$, as they have nothing to do with human intentionality.

However, the distinctions referred to should not be understood in dualistic terms. Traditionally, following Descartes, dualism involves a masterclass and an underclass. In the mind/body dualism, mind is superior to body; in the male/female dualism, male is superior to female (see Plumwood 1993). But in the context of this discussion, humans are not meant to be superior to nonhumans; neither is human culture superior to nature. Humankind is neither superior nor inferior to the rest of nonhuman nature, just unique in terms of its peculiar consciousness in a way analogous to the uniqueness of the beaver with its ability to build its dams. Far from regarding nature$_{fa}$ to be inferior, the nonanthropocentric perspective adopted here advocates that we, humans, should be sensitive to the status of nature$_{fa}$ as "ontological other." Should we systematically undermine nature$_{fa}$, we would be creating an ontologically diminished world, indeed, a narcissistic civilization in which everything we see, hear, touch, smell, and move about is but our own handiwork embodying our human, not nature's own, creativity, complexity, or ingenuity.

NOTES

1. The definition of ontology and the ontological objection to the loss of naturally occurring forms of biodiversity will be made clear in the body of the chapter.

2. As we shall see in a later section, they are extremely powerful, especially in the political context of trying to save or to study biodiversity. Furthermore, according to the dominant strand of Western philosophy even today, only instrumental reasons are intelligible.

   Such reasons necessarily fall within an anthropocentric framework (which has as its axiom the postulate that humankind alone is intrinsically valuable), and may be classified under two headings. Biodiversity loss either causes harm directly to humans, or indirectly to humans via direct harm to nonhumans. An obvious example of the former would be the extinction of a plant which humans might find nutritious. An example of the latter would be the loss of individual organisms and threat to species when their habitat is threatened by, say, oil pollution. The pollution directly harms nonhuman others, but as these individual organisms and their species may be specially valued by humans, the pollution thereby indirectly harms humans. However, according to this perspective, to recognize direct harm to nonhumans does not mean that we humans have a direct moral duty to them in preventing that harm. On the contrary, it recognizes direct duties to humans alone and only indirect duties to nonhumans. For instance, if a neighbor kills your pig, he has failed in his direct duty not to cause you (in this case, economic) harm; he has no direct duty not to harm the pig itself, only an indirect duty via direct duty to fellow humans.

3. For further discussion of the limitations of the biological species concept and the difficulties about ascertaining the number of species, see Lee 1998.

4. Naturally occurring biodiversity, as species diversity, may in certain contexts even be increased through active management and rapid-focused intervention; see, for instance, Anderson and Devlin 1990.

5. For a more detailed account of the history of technology as well as its relationship to the history of science, the philosophy of science, and the philosophy of technology, see Lee 1999, chap. 2; for a detailed account of the relationship between Mendelian genetics and double-cross hybridization on the one hand and DNA genetics and biotechnology (as genetic engineering) on the other, from the point of view of the history of science, see Lee 2002.

6. Distinctions between phenotype and genotype on the one hand and dominant and recessive genes on the other make it absolutely clear (a) why an organism which does not manifest a certain characteristic (encoded by the recessive allele or version of the gene) may nevertheless be carrying the genetic material in its genotype; (b) that such genetic material may be recovered in the phenotype of a particular descendant under certain specific circumstances. Thus, according to the law of segregation, although all first (F1) generation offspring phenotypically express the dominant gene, they are, however, in their genotypes, hybrids (Aa, Aa, Aa, Aa) having

169

received dominant alleles from one parent (AA) and recessive ones from the other (aa). In the second (F2) generation, when two such hybrids are crossed (Aa with Aa), their genotypes pan out as AA, Aa, Aa, aa. Phenotypically, the ratio of dominant to recessive characteristics is 3:1; but genotypically, two are hybrids (Aa), and two are not, with one of these containing two alleles for the dominant gene (AA) and the other, two alleles for the recessive gene (aa) – in other words, there are three different genotypes.

7. In this context, "anthropogenic" simply means that humans are the origin or cause (of something); "nonanthropogenic" means that the source or cause (of something) has nothing to do with humans and their activities.

8. Already, the major biotech companies, like Monsanto, are arguing that the world's hungry in the twenty-first century would and could not be fed unless experiments in genetically modified foods were to go ahead unhindered by governmental interference. (This, of course, ignores the counterargument that the hungry of the world today go hungry because they are poor, rather than because of the actual shortage of food grown and being available, in theory, for consumption.)

9. The average age of a mammalian species is roughly a million years; modern *Homo sapiens* has lived about 100,000 years.

10. Other nonanthropocentric environmental philosophers have, in various ways, vociferously challenged the thesis that humans are the sole locus of intrinsic value.

11. Ontology is that branch of philosophy which is concerned with being, and what kinds of being there may exist. For instance, God (or gods/goddesses, angels) is a very different kind of being from that of plants and animals (including humans in this context). The former involves supernatural beings while the latter, in contrast, are said to be natural beings. Natural beings (in this sense of "natural" which is the antonym of "supernatural") are physical beings with space-time coordinates; supernatural beings, *ex hypothesi*, are outside the physical domain.

   This contribution is specifically concerned to argue for the ontological distinction (or dyadism) between the natural and the artefactual, a distinction which will be made clear in what is to follow.

12. There are, obviously, other senses of the term nature which will not be considered here. For a detailed clarification of the different senses, see Lee 1999.

13. The term trajectory is introduced here as a technical one, to refer to the entire history of a naturally occurring item, whether biotic or abiotic. For further details, see Lee 1999.

14. In accordance with such an argument, we might as well claim, too, that nemotodes are three-quarters human as they share 75 percent of their DNA with us.

15. However, intellectual capability constitutes only a necessary but not a necessary and sufficient condition for moral agency.

16. Philosophical issues should not be confused with purely verbal or terminological matters. If someone feels that the words "primary" and "secondary" are not the appropriate ones to use, then they may be replaced by some other terms, perhaps, like "underlying" and "what lie on the surface."

17. However, to say that a biotic artefact is every bit as much an artefact as an abiotic artefact is not to deny that both biotic and abiotic artefacts may possess different levels or depth of artefacticity. For instance, a traditionally domesticated cow displays artefacticity at a lesser depth or level than the transgenic cow. The former

is the result of crossing a male and female belonging to different varieties of the species, whereas the latter is the result of crossing species, and indeed, even kingdom barriers.

18. The use of Aristotle's four causes is limited in this discussion, confined only to the light they may be able to cast on the notion of artefact. As a complete explanatory schema, modern science and modern philosophy have abandoned it, retaining only two of the four causes, namely, material and efficient.

19. See, for instance, Easterbrook 1995. Indeed, Easterbrook rightly maintains that anthropogenic loss of biodiversity could be said even to be puny compared to nonanthropogenic loss in the past.

BIBLIOGRAPHY

Anderson, G. T. and C. M. Devlin. 1990. Restoration of a Multi-species Seabird Colony. *Biological Conservation* 90:175–81.

Easterbrook, G. 1995. *A Moment on the Earth: The Coming Age of Environmental Optimism*. London: Penguin Books.

Lee, K. 1998. Biodiversity. In *Encyclopedia of Applied Ethics, Volume 1*. R. Chadwick, ed. San Diego: Academic Press.

   1999. *The Natural and the Artefactual: Implications of Deep Science and Deep Technology for Environmental Philosophy*. Lanham: Lexington Books.

   2002. *Philosophy and Revolutions in Genetics: Deep Science and Deep Technology*. London: Palgrave.

Plumwood, V. 1993. *Feminism and the Mastery of Nature*. London: Routledge.

Takacs, D. 1996. *The Idea of Biodiversity: Philosophies of Paradise*. Baltimore and London: John Hopkins University Press.

Varner, G. E. 1998. *In Nature's Interests?: Interests, Animal Rights and Environmental Ethics*. New York and Oxford: Oxford University Press.

8

Evaluating Biodiversity for Conservation

A Victim of the Traditional Paradigm

PETER R. HOBSON AND JED BULTITUDE

The conservation strategy for nature reserves in modified landscapes in the United Kingdom is essentially based on the "historical principle" (Peterken 1996) which maintains the argument that wildlife will be best served by continuing the historic form of land use practice in those ancient seminatural habitats that have a past record of management.

The countryside across much of Europe has been shaped by centuries of intensive management that has given rise to a cultural landscape (Fry 1991) characterized by features which may include coppice woodlands, heathlands, grazed flood meadows, and organized reed beds. According to Morris, in the United Kingdom many of these areas represent the cherished landscape and are designated as nature reserves, parks, or other forms of protected land. An important element in landscape and wildlife conservation has been the appeal to historical precedent. Consequently, to the modern conservationist, species of young sere-stage communities come to represent the norm. Furthermore, past management practices, which have shaped this natural heritage, are now enjoying a renaissance as the modern solution to conserving biodiversity by maintaining the status quo. The biological interest of many of the reserve sites shows that its past management has been successful, up to a point (Morris 1991). However, we know almost nothing of what has been lost from biologically rich sites as our basis for comparison is contemporary. The virtual absence of habitat existing in a state of "original-naturalness" (Peterken 1981) in Europe has made the task of constructing a model of a "natural" ecosystem almost impossible.

172

8.2 THE TRADITIONAL PARADIGM, 'RECEIVED WISDOM'

In the United Kingdom the broader conservation rationale for the preferred choice of traditional management includes the following reasons:

- It perpetuates existing native community types and generates diverse habitats on the assumption that if a species has survived traditional management so far, it will continue to survive under that treatment in the future.
- It maintains the character of the historically modified landscape and in so doing sustains values recognized in that particular system.

These two standpoints uphold the principle of the traditional paradigm, which perceives product-based biodiversity as the central focus to conservation, "devise actions which will deliver the objectives and targets within a given timescale" (Wynne et al. 1995). The more rational arguments that traditional land management was never designed to protect diversity but to maximize agricultural, silvicultural, or industrial output, and that many species survived *despite* past practices (Hambler and Speight 1995), are often not addressed by traditionalists.

More recently in Europe conservation has undergone a fundamental change following the signing of the Biodiversity Convention at the United Nations Conference on Environment and Development in Rio de Janeiro in 1992. Specifically in the United Kingdom, a long-term strategy for conserving biodiversity has been formulated which includes identifying priorities for action with clearly defined targets. Furthermore, detailed action plans are to be produced for habitats and species with quantifiable targets, which will make it possible to monitor the overall success of the plan. To aid the process of biodiversity action planning it has been necessary to compile from existing data a conservation priority listing for species or larger taxonomic groups considered to be rare or endangered. Finally, by studying the known distribution of priority species, it has been possible to relate these patterns to the broader ones of land use, and so identify priority landscapes and their associated management practices for preservation. For many species and habitats this has meant a marriage of the principles and practices of both the traditional and equilibrium paradigms. Unfortunately, this relatively new science is based on contemporary patterns of species distribution and landscape forms and as such is unable to shed light on the true ecological identity of the region before modification took place. It is conceivable that much of the wildlife we prioritize for conservation is well equipped with survival strategies that have enabled it to withstand many centuries of habitat modification and so persist despite change (Hambler and Speight 1995). If this is the case, then we need to question the whole process of using priority species as indicators of the health and stability of naturally functioning ecosystems.

8.3   ASSESSING AND INTERPRETING BIODIVERSITY

The task of conserving wildlife has been made more effective by developing a convenient hierarchical classification and measurement system which categorizes nature as discrete entities at a range of scales from genetic and taxonomic categories to habitat and landscape units (Hunter 1996; Spellerberg and Sawyer 1999). This process has enabled scientists to systematically record, differentiate between, and rank wildlife as well as monitor any changes. The perception of nature as existing in this unitized form has made it possible for scientists to detect downward trends in the diversity of species and habitats – compelling evidence to persuade policy makers that swift action is required. Equally, any upward trend in target indicators recorded for specific sites has been interpreted as a measure of the success of a particular conservation program. This form of assessing nature provides a strong rationale for the implementation of management strategies that have as their prime objective to increase and sustain numbers of species, communities, and habitats in a given area. Often the language adopted in conservation is reminiscent of business-speak, "increase numbers of target species to new stated levels within the natural range of the species" (Wynne et al. 1995). Species-action plans talk of mission statements, strategic objectives, operational objectives, prescriptions, and wildlife gain. "The UK biodiversity should be sharply focused on the outcome – what needs to be achieved for individual species, biodiversity challenge is objective-led" (Wynne et al. 1995). This approach to conservation gives the impression that wildlife is perceived as a commodity and as a consequence is exposed to value judgments of redundancy, decline, and awarded status. Furthermore, it provides many conservationists with a convenient and persuasive means of evaluating the success of active management (Wynne et al. 1995).

In the United Kingdom, over the last thirty years, a large number of sites have been prioritized for conservation management. The site selection procedure has involved an evaluation process that has included in its list of key criteria species diversity, fragility, typicalness, rarity, naturalness, ecological integrity, potential, and intrinsic appeal (Ratcliffe 1977). Like so many evaluative systems, these criteria are based on qualitative assessments. Such surveys are characterized by the transparency of their underpinning values

(Usher 1986; Spellerberg 1991). Despite these problems, these criteria, or variations of them, are still in current use. The criterion – the "potential" – of a site is often defined in terms that are strongly suggestive of a conservation strategy heavily reliant on a policy of active intervention and management to achieve results, "reinstatement of coppice management; particularly in recently neglected sites" (Wynne et al. 1995).

The more recent use of phytosociological classification in the assessment of biodiversity has given conservation managers the opportunity to describe the diversity of nature in another form and at a different scale to that of just population or species. It is now possible to compartmentalize natural and seminatural landscapes into classifiable categories. There now exists for many parts of Europe a detailed record of the wide range of essentially seminatural vegetation types which forms the basis for the Annex I habitats listing (priority habitats for conservation as listed under the European Habitats Directive, 1992). This form of nature assessment is designed to emphasize the abundance of discernible patterns in the composition of vegetation as a premise for biodiversity rather than biological relationships, functional roles, and dynamics. Recent developments in conservation science which address the ecological importance of open ecosystems and of ecosystem processes and functions (Christensen 1997; Tartowski et al. 1997; Hunter 1999) have yet to be translated into important criteria for the purpose of evaluating biodiversity for conservation. For instance, consider an evaluation system which does not measure quality by hierarchical classification of natural attributes, but rather one which measures the threat to the sustainability of natural functioning ecosystems. However, there is little evidence that this form of evaluation is in use; rather, new evaluation techniques assess biodiversity on the basis of a "species-based reserve approach" (Gray 2000).

8.4 MAINTAINING THE STATUS QUO – THE DESIRED STATE PRINCIPLE

The sophisticated process of describing, classifying, assessing, and evaluating our wildlife assets has set the conservationists on a course of devising elaborate management plans and targets for preserving in state these atomized attributes of nature. Such a strategy may be described as the "desired state principle" and can be summarized as follows: the continuation of an inheritance of traditional management practices or specific species-habitat management which have created the character of old landscapes and also preserved a distinctive wildlife community.

Consequently, there is a need to:

- maintain a selection of prioritized habitats and species within "specified limits of acceptable change" (Sutherland and Hill 1995) as nature, in the current situation, is no longer able to manage the system;
- manage habitats in order to conserve representative units of vegetation as described by phytosociological classification systems of seminatural landscapes;
- manipulate or maintain natural systems to sustain cultural and wildlife values recognized in those systems (Fiedler et al. 1997);
- maintain and where possible increase populations of "flagship" and "charismatic" species as they serve as a barometer to the quality of a habitat;
- ensure that the "inherited" diversity of existing seminatural habitats is maintained or "enhanced."

The emphasis is very much on prioritizing and preserving a diverse range of representative seminatural types as subsets of larger landscape units (Barrett and Barrett 1997). Natural processes, catastrophes, or stochastic events are perceived as a greater threat to wildlife habitats than anthropogenically in-duced environmental stochasticity (Wiens 1997).

These values are translated in a typically conventional or prescriptive way. For instance, in British woodlands, late successional species such as Small leaf lime (*Tilia cordata*) and Hornbeam (*Carpinus betula*) are maintained in their traditional coppice form because of the value attached to a recognizable assemblage of vernal plants which benefit from this type of woodland struc-ture: "Normally, recoppicing is the best conservation policy; [to preserve] bluebell woods" (Rackham 1986). In many cases, certain species including Hazel dormouse (*Muscardinus avellanarius*) and even nonthreatened species such as Badger (*Meles meles*) achieve emblem or high intrinsic appeal sta-tus and consequently are integrated into management objectives for certain woodlands. If traditional woodland practices are judged to be beneficial to specific conservation species such as Hazel dormouse in the United Kingdom, then it is natural to assume that the broader ecological interest of the habi-tat is also protected. However, many conspicuous woodland species in the United Kingdom valued by conservationists are typical of an open habitat and consequently any form of management which promoted a high diversity of open habitat may lead to an impression of high species richness (Hambler and Speight 1995) irrespective of the quality of the ecosystem. The more "retiring" species that may have an important functional role are generally overlooked (Whitbread and Jenman 1995). As a consequence of this philoso-phy, many woodlands are managed specifically to ensure that certain desired

characteristics are promoted and valued species are promoted to favorable
levels of abundance as set by targets.

8.5   SHIFTING PARADIGMS – ADOPTING THE SCIENTIFIC
PARADIGM IN CONSERVATION

In recent years, advances in ecological science and understanding have
brought about a shift in the way biodiversity is perceived and valued from
the conventional "equilibrium worldview" to a more informed "nonequilibrium worldview" (Fiedler et al. 1997). This paradigm shift has encouraged
scientists to reexamine fundamental ecological principles related to the conservation of biodiversity. These include the need to protect the integrity of
ecosystem function, systems, and heterogeneity – to "gain knowledge and understanding of the contents, structures and functions of the ecosystems of the
area and region, to support decision making, management and use" (Miller
1994). In the past, traditional approaches to conservation have focused on
valued objects of a nation's ecological heritage, such as rare and endangered
species, critical habitats, and unique communities (Barrett and Barrett 1997).
In the United Kingdom, Peterken (1996) suggests that three management
strategies are practiced in the conservation of seminatural woodlands: management to promote natural function and structure of woodlands; traditional
or precautionary management which yield higher species or habitat diversity;
and design management to meet specific targets such as preserving rare or
endangered species. This approach is brought into sharp contrast with the
strategies adopted by the United States and Scandinavia by its failure to recognize the inseparable relationships between species ecology and ecosystem
dynamics. The concept of "close to nature" management is not a new one
and in silviculture it has been applied for more than a hundred years (Schütz
1999). However, even with this relatively liberal and pragmatic philosophy
there is a broad spectrum of interpretation between two diametrically opposing paradigms, cultural and natural. This not withstanding, the overriding
principle is "selfing" systems. Self-functioning systems have a chance of sustaining themselves and ensuring that biodiversity is preserved. In both North
America and Scandinavia the principle of biodiversity as a function of natural forest dynamics is widely accepted in forestry management. For instance,
action plans for conservation proposed by the National Board of Forestry
in Sweden (1996) accept this principle implicitly. Also, in both the United
States and Canada, the "ecosystem management" approach underpins forest
management policy (Kimmins 1997).

Conservation has been slow to adopt these recent developments in ecology, and for many the viewpoint that any object or unit of nature of ecological interest is considered conservable because natural systems are still perceived to be static, closed, and fixed holds true. This is evident in the systems of assessment and evaluation for biodiversity.

8.6 CONCLUSIONS

The important challenge ahead for conservationists is to devise a holistic approach to conserving woodlands, which recognizes the relationship between diversity of form, ecosystem function, and sustainability. A shift away from an essentially equilibrium paradigm toward a more nonequilibrium paradigm approach to conservation management would require a change in the values and priorities afforded to different attributes of nature (Barrett and Barrett 1997). Essentially, conservationists would need to put into practice the principle that valued objects of a nation's ecological heritage such as rare species, communities, and habitats are now perceived as integral components of open, heterogeneous, and dynamic ecosystems (Barrett and Barrett 1997). Ultimately this would impact both the assessment and evaluation processes as well as the final management of wildlife.

BIBLIOGRAPHY

Barrett, N. E. and J. P. Barrett. 1997. Reserve Design and the New Conservation Theory. In *The Ecological Basis of Conservation*, S. T. A. Pickett, R. S. Ostfeld, M. Shachak, and G. E. Likens, eds. New York: Chapman and Hall.

Christensen, N. L. 1997. Managing for Heterogeneity and Complexity on Dynamic Landscapes. In *The Ecological Basis of Conservation*, S. T. A. Pickett, R. S. Ostfeld, M. Shachak, and G. E. Likens, eds. New York: Chapman and Hall.

Fiedler, P. G. and P. M. Kareiva. 1998. *Conservation Biology: For the Coming Decade*. New York: Chapman and Hall.

Fiedler, P. G., P. S. White, and R. A. Leidy. 1997. The Paradigm Shift in Ecology and Its Implications. In *The Ecological Basis of Conservation*, S. T. A. Pickett, R. S. Ostfeld, M. Shachak, and G. E. Likens, eds. New York: Chapman and Hall.

Fry, G. L. A. 1991. Conservation in Agricultural Ecosystems. In *The Scientific Management of Temperate Communities for Conservation*. I. F. Spellerberg, F. B. Goldsmith, and M. G. Morris, eds. Oxford: Blackwell Science.

Gray, A. N. 2000. Adaptive Ecosystem Management in the Pacific Northwest: A Case Study from Coastal Oregon. *Conservation Ecology* 4(2): 6.

Hambler, C. and M. R. Speight. 1995. Biodiversity Conservation in Britain: Science Replacing Tradition. *British Wildlife* 6.3:137–47.

Hunter, M. L. 1996. *Fundamentals of Conservation Biology*. Oxford: Blackwell Science.

1999. *Maintaining Biodiversity in Forest Ecosystems*. Cambridge: Cambridge University Press.

Kimmins, J. P. 1997. *Forest Ecology. A Foundation for Sustainable Management*. Englewood Cliffs, Prentice Hall.

Miller, K. R. 1994. International Cooperation in Conserving Biological Diversity: A World Strategy, International Convention, and Framework for Action. *Biodiversity and Conservation* 3:464–72.

Morris, M. G. 1991. The Management of Reserves and Protected Areas. In *The Scientific Management of Temperate Communities for Conservation*, I. F. Spellerberg, F. B. Goldsmith, and M. G. Morris, eds. Oxford: Blackwell Science.

Peterken, G. F. 1981. *Woodland Conservation and Management*. London: Chapman and Hall.

1996. *Natural Woodland: Ecology and Conservation in Northern Temperate Regions*. Cambridge: Cambridge University Press.

Rackham, O. 1986. *The History of the Countryside*. London: Phoenix Giant.

Ratcliffe, D. A., ed. 1977. *A Nature Conservation Review*, 2 vols. Cambridge: Cambridge University Press.

Schütz, J.-Ph. 1999. Close-to-Nature Silviculture: Is This Concept Compatible with Species Diversity? *Forestry* 72.4:359–66.

Spellerberg, I. F. 1991. *Monitoring Ecological Change*. Cambridge: Cambridge University Press.

1992. *Evaluation and Assessment for Conservation*. London: Chapman and Hall.

Spellerberg, I. F. and J. W. D. Sawyer. 1999. *An Introduction to Applied Biogeography*. Cambridge: Cambridge University Press.

Sutherland, W. J. and D. A. Hill. 1995. *Managing Habitats for Conservation*. Cambridge: Cambridge University Press.

Tartowski, S. L., E. B. Allen, N. E. Barrett, A. R. Berkowitz, R. K. Colwell, P. M. Groffman, J. Harte, H. P. Possingham, C. M. Pringle, D. L. Strayer, and C. R. Tracy. 1997. Toward a Resolution of Conflicting Paradigms. In *The Ecological Basis of Conservation*, S. T. A. Pickett, R. S. Ostfeld, M. Shachak, and G. E. Likens, eds. New York: Chapman and Hall.

Usher, M. B., ed. 1986. *Wildlife Conservation Evaluation*. London: Chapman and Hall.

Whitbread, A. and W. Jenman, 1995. A Natural Method of Conserving Biodiversity in Britain. *British Wildlife* 7.2:84–93.

Wiens, J. A. 1997. The Emerging Role of Patchiness in Conservation Biology. In *The Ecological Basis of Conservation*, S. T. A. Pickett, R. S. Ostfeld, M. Shachak, and G. E. Likens, eds. New York: Chapman and Hall.

Wynne, G., M. Avery, L. Campbell, S. Gubbay, S. Hawkswell, T. Juniper, M. King, P. Newbery, J. Smart, C. Steel, T. Stones, A. Stubbs, J. Taylor, C. Tydeman, and R. Wynde. 1995. *Biodiversity Challenge*, 2d ed. Sandy: Royal Society for the Protection of Birds.

9

Limits to Substitutability in Nature Conservation

DIETER BIRNBACHER

9.1 BIODIVERSITY AND THE "SUBSTITUTION PROBLEM"

Philosophy traditionally deals with concepts and values, either in the spirit of analysis and reconstruction or in the spirit of construction and innovation. In both respects, biodiversity is a "hard case" and a challenge to philosophy's intellectual resources. Like simplicity (a concept which plays a crucial role in the philosophy of science and scientific methodology), diversity is a concept of considerable complexity which easily defies definition. With biodiversity, difficulties multiply because it is not clear what the items are of which biodiversity is predicated. What does it mean to say that a natural system is "diverse," absolutely or to a certain degree? Does diversity refer to the number and diversity of biological species, to the number and diversity of alleles, or to the diversity of all properties of natural systems including structural, aesthetic, and symbolic properties? The distinction, customary in ecology, between species diversity, ecosystem diversity, and genetic diversity is certainly helpful in this respect, since it shows the essential incompleteness of statements about biodiversity. Nevertheless, an inclusive concept of biodiversity, covering all relevant dimensions, would certainly be desirable.

Similar difficulties beset the attempt to spell out why biodiversity, however defined, is a value and why we should be concerned to preserve or even enhance biodiversity. Is the value of biodiversity extrinsic or intrinsic? "Extrinsic" means that it is valued for its causal role in preserving or enhancing other values, "intrinsic" that it is an end in itself worth pursuing for its own sake. If its value is solely or predominantly extrinsic, what are the values it is meant to serve and what is the status of these values? Are they anthropocentric (such as long-term economic well-being, aesthetics, and education) or biocentric (such as existence, plenitude, or nonhuman nature's chance to evolve undisturbed by human intervention)?

In the following, attention is drawn to another difficulty surrounding bio-diversity as a value concept – its indifference to *individuality*. Biodiversity, taken as a central value concept of nature ethics, confronts a problem iden-tified by Eric Katz as the "substitution problem" (Katz 1985). The problem is that, in principle, every entity can be replaced by another entity possessing the same valuable properties. Provided that the overall value of nature is not diminished, or even increased, any replacement of one entity by another is legitimate, or even required. "Overall value" means here the aggregate of the intrinsic and extrinsic value of the natural entity in question, ascribed on the basis of a certain underlying axiology and a causal analysis of its short- and long-term effects.

The original context of Katz's diagnosis of the "substitution problem" was that of ecological *holism*: According to holism, only natural wholes such as biotopes, ecosystems, or biological species are admitted as bearers of nat-ural value. This implies that individual natural entities have value only *qua* components of the wholes of which they are parts. Except for cases in which the value of a natural whole partly derives from the identity of some of its components (as with natural monuments, for example), every single compo-nent of the valuable whole is in principle replaceable by another individual with the same function or role. Thus, the value of a forest does not in general depend on the individuality of the trees of which it is made up. It follows that even massive changes in an ecosystem by human intervention can be justified given that the overall value of the relevant whole remains constant or is increased, for example by increasing its diversity, complexity, interre-latedness, or aesthetic attractiveness. If *function* or *role* is all that matters, any entity can in principle be replaced by another with the same function. If, in Aldo Leopold's famous dictum "a thing is right when it tends to pre-serve the integrity, stability, and beauty of the biotic community" (Leopold 1949, 224f.) "integrity" is interpreted as the maintenance of the system in its *qualitative* aspects, any substitution of natural entities by human interven-tion is permitted which maintains or even adds to the integrity of the biotic community in this sense. The logical basis of substitution is, in this case, the simple ontological truth that the identity of a *whole* of some sort is in general compatible with the destruction, or loss, of one or the other of its individual components. While maintaining the integrity of an *individual* entity implies assuring its survival as an individual, maintaining the integrity of a *system, organic unity*, or *whole* can be taken to imply no more than the maintenance of its essential properties, its defining qualitative aspects. (This is, however, only one possible interpretation of Leopold's "integrity." It can also be taken to mean inviolability in a stricter sense, prohibiting destruction and substitution of its individual components.)

As far as the preservation of biodiversity is concerned, the consequences are evident. Biodiversity as a second-order holistic value (a value attaching to a whole of wholes) does not imply the sacrosanctness of any individual whole (genome, ecosystem, or species) that contributes to it. Since only function matters, any individual is in principle replaceable. If restrictions of substitutability are in place, these are due to *contingent* factors, such as the slowness of biological evolution. As Jakobson and Dragun (1996, 60) state, "the genetic impoverishment caused by present extinction rates will not be replenished for about 5 million years." This fact may well afford a reason to protect each species individually to secure species diversity in the long run.

Katz interpreted the "substitution problem" as a problem peculiar to holistic conceptions of natural value. In fact, holism is only a special case. The problem is more general and arises in the same way for nonholistic conceptions. In *holistic* axiological models, individuals are substitutable under their *functions* for valuable wholes. In *individualistic* axiological models, individuals are substitutable under their value-conferring *properties*. Substitutability is not avoided by substituting holistic for individual value – as Katz seems to think when he postulates that "an environmentally conscious moral decision maker cannot merely consider the instrumental value, the functional operation of the entities in the ecological system; he must also consider the integrity and identity of the entities in the system, i.e. their intrinsic value" (Katz 1985, 255). If "intrinsic value" in this quotation is interpreted in the usual way, that is, as a value ascribed to a thing in virtue of its own properties (in contrast to extrinsic value ascribed in virtue of its effects), individual intrinsic value (exceptions apart) fares no better, with respect to substitutability, than holistic intrinsic value. However intrinsic value is conceived, as beauty, richness, complexity, or whatever, it allows for substitution of individuals in the same way for ecological individualism as for ecological holism. Each of these values is preserved by substituting one individual for another with the same value. Even in the *community* model which Katz proposes and which takes into account *both* the instrumental value of the individual entity for a larger organic whole *and* the intrinsic value of the individual, substitutability is preserved. If I substitute, in my garden, a beautiful forsythia by an even more beautiful jessamine, without thereby infringing on the functioning of my garden as a biotope, I certainly *increase* rather than *diminish* the overall value of my garden, as far as its intrinsic value is concerned.

9.2   THE 'SUBSTITUTION PROBLEM' AND AXIOLOGICAL APPROACHES TO NATURE ETHICS

Axiological or value ethics bases human obligation on the real or potential existence of entities exhibiting certain value properties. The basic moral obligation postulated by axiological ethics is either to assure the maintenance of a given level of value-instantiation (in a certain domain), or to increase the level of value-instantiation (in a certain domain), depending on whether the ethical system takes a more conservative or a more progressive turn. (An essentially *conservative* norm of maintaining a certain level of value-instantiation is part and parcel of the idea of *sustainability*.)

Axiological approaches continue to dominate nature ethics, but there are also a few rival theories. In Germany, *deontological* positions play a more prominent role than in the Anglo-Saxon world, such as Albert Schweitzer's ethic of *reverence of life*, and Kantian approaches basing obligations toward animals and other components of nonhuman nature on the concept of *human dignity*. A more recent, and prima facie more plausible, alternative to axiological approaches is the *genetic* approach suggested by biologist Kattmann (1997). According to Kattmann, obligations in respect of the natural world do not arise from the *values* embodied in nature but from our position within the natural world. Responsibility for nature is based not on any qualitative properties of natural individuals or systems but on relations like descent and kinship. We are responsible for nature not in the way a custodian is responsible for treasures valued for their beauty, rarity, or old age, but in the way a father is responsible for his children, which is unconditionally and irrespective of any specific value attributed to them by himself or others. Ecological responsibility is dictated by fact and insofar beyond dispute and human discretion.

Many of these rival approaches to nature ethics are inspired by the very same observation which underlies the "substitution problem": that axiological approaches primarily attach value to general *properties* (qualities, functions, relations) exhibited by natural entities and not to these *individuals* themselves. They attach value to individual natural entities *qua* bearers of certain properties but are ultimately indifferent to the existence of these individuals themselves. In this way, they are at odds with one of the central impulses of nature conservation, the impulse to preserve natural entities for their own

183

sake, in view of their singular individuality, or, to use a rather old-fashioned expression, their *haecceitas*.

Values are treated by axiological approaches as being of *adjectival* character, as *supervenient* on certain descriptive properties. If an individual natural entity is intrinsically worthy of protection, this is, according to axiological theories, in virtue of some general property the entity has. It may, for example, be worthy of protection because it adds to the overall level of beauty, complexity, and stability of an ecosystem. However, most values for which it is protected under the axiological scheme have no relation to its concrete individuality. If, for example, an entity is protected because it adds to the overall level of biodiversity, the value for which it is protected has nothing to do with any of its individual properties. The only property by which it contributes to biodiversity is that of being sufficiently *different*, in genotype, phenotype, or some other relevant property, from other natural entities.

How serious is the "substitution problem"? Does it seriously weaken the plausibility of axiological approaches to nature ethics (and to biodiversity protection in particular), or can it somehow be reconciled with these approaches?

The first thing that should be said is that the extent to which natural entities are made substitutable by an axiology of natural values depends on the particular axiology chosen and on how this axiology is organized: whether it is *monistic* or *pluralistic*, on what level of *generality* its values are defined, and whether these values are *commensurate* or not, allowing or not allowing for an aggregation of different kinds of value into one overall value. If the values of a pluralistic axiology are taken to be commensurate, we get substitutability *among* values in addition to substitutability under one single value. Trade-offs become possible between the degrees in which different values are realized by a natural entity, so that entity e_1, which realizes values v_1, v_2, and v_3 to degrees d_1, d_2, and d_3, respectively, will be replaceable by an entity e_2 which realizes values v_4, v_5, and v_6 to degrees d_4, d_5, and d_6, respectively. Even if jessamine, in my personal axiology, is less beautiful than forsythia, jessamine may make up for this by attracting, and supporting, a far greater variety of butterflies, thus realizing the same overall value. Not only is the beauty of e_1 traded off against the beauty of e_2, but beauty is traded off against diversity. If the values concerned are incommensurate, the range of substitutability among values, and thereby of entities bearing different kinds of value, is considerably narrowed. If values v_1 to v_5 are lexicographically ordered with v_1 (beauty, say) at the top, substitution of jessamine for forsythia would diminish the overall value of my garden, no matter what other relevant differences there are, and would be incompatible with even the weakest normative principle of an axiologic ecological ethics, the principle of value preservation.

It may be questioned whether talk of substitution and substitutability makes sense at this high level of aggregation. That e_1 is replaceable by e_2 always means that e_1 is replaceable by e_2 *in a certain respect r*. In our case, r is a *value* aspect, a "covering value" (Chang 1997, 5 ff.). But can r be an abstract value without any specific content? It makes perfect sense to say that e_1 is replaceable by e_2 with respect to beauty or diversity (which is to say that e_1 is more beautiful or diverse than e_2). But does it make sense to say that e_1 is substitutable by e_2 with respect to value or valuableness, without specifying the concrete nature of this value?

I do not think that the abstractness of the value aspect in this case is incompatible with meaningful talk of substitution and substitutability. Even in cases in which the overall value of a state of affairs is no specific value, but the aggregate of all value aspects present in it (for which we have no specific label apart perhaps from "desirability"), it seems a perfectly legitimate aspect with respect to which two states of affairs can be equivalent or nonequivalent, substitutable or nonsubstitutable. The situation may be compared to cases in which we judge that a certain distribution is *just*, in an overall sense, where justice is the aggregate of a number of partial aspects of justice (like fairness, equality, differentiation according to merit, need, etc.). We do compare and evaluate distributions in terms of overall justice, despite the fact that this measure cannot be explicated but by reference to a diversity of criteria or principles of justice.

9.3 WHY IS SUBSTITUTION A 'PROBLEM'?

Substitutability is a persistent issue in the controversy between economists on the one hand, and environmentalists and several schools of environmental ethics on the other. For economists, substitutability of natural entities tends to be the rule, for environmentalists it tends to be the exception. For economics, substitution is a normality. For a large fraction of environmentalists, and for many practicing ecologists, substitution is a provocation.

Explanations are not hard to find. Economists typically favor *anthropocentric* axiologies, whereas environmentalists typically do not. "Efficiency" in economics usually means the maximization, or satisficing, of *human* preferences. Since in an anthropocentric axiology the value of nonhuman natural entities can only be *extrinsic* value, in other words, *resource* value in the widest possible sense (including aesthetic, scientific, spiritual, and other kinds of "inherent" anthropocentric value in Frankena's sense [Frankena 1979, 13], it is evident that bearers of extrinsic values are exchangeable by their very nature. A mere means is always substitutable by functional equivalents. If natural entities are viewed as means for human ends, or as raw material to be transformed, in one way or other, into human goods, the relation between nonhuman nature and value is *contingent*. Nothing depends on the exact nature of the natural inputs into the human states and activities in which intrinsic value resides.

Another reason is that quite a number of economists subscribe to some variant of *welfarism* (such as utilitarianism) or to some other *monistic* value system, whereas environmentalists typically favor pluralistic value systems, often with lexicographic orderings. Value monism implies substitutability, whereas value pluralism does not. If there are several distinct values, these need not have a common denominator.

Third, it is obvious that the resources of economics are much better suited to problems allowing unlimited substitution. There is a natural convergence of what the mathematical machinery of economics is most apt to deal with and the intellectual predilections of those who wield this machinery. Environmentalists, on the other hand, tend to have an almost instinctive dislike of formal approaches, sometimes misleading them into a full-blown rejection of formal methods of systematizing values, establishing priorities, and calculating costs and benefits.

Environmentalists tend to reject substitutability of natural entities, though mostly without any clear theory about what they are doing. The fact that the majority of environmentalists are in favor of nonanthropocentric and non-welfarist axiologies and subscribe to some form of axiological pluralism is, as we have seen, not sufficient to explain the rejection of substitutability. Even granting that there are problems in any *practical* substitution of natural entities, for example because of possible side effects, risks of irreversibility, and limited possibilities of creating new species or alleles, all this cannot justify the strictness with which substitution is rejected in some strands of ecological ethics even as a thought experiment. David Ehrenfeld, for example, writes as if each natural value is in principle incommensurate with any other so that the condition of equivalence basic to substitutability is never fulfilled:

> The . . . hazard is that formal ranking is likely to set Nature against Nature in an unacceptable and totally unnecessary way. Will we one day be asked to choose between the Big Thicket of Texas and the Palo Verde Canyon on bases

of relative point totals? The need to conserve a particular community or species must be judged independently of the need to conserve anything else. Limited resources may force us to make choices against our wills. But ranking systems encourage and rationalise the making of choices. (Ehrenfeld 1978, 203ff.)

Edward O. Wilson expresses a similar conviction by saying in a newspaper article that "each biological species is a masterwork of evolution and irreplaceable" (1995, 33). Though the exact meaning of "irreplaceable" is far from clear in this statement, it is, I think, obvious that it warns us not to deal with natural entities (and especially natural species) in the calculating and actuarial spirit characteristic of economics.

Is it possible to give a rational account of this rejection of substitution and substitutability? What is the reasoning behind this wholesale opposition to any principled weighing, comparing, and prioritizing of values, the part and parcel not only of economics but of axiological thinking in general?

One potential answer is that the assumption of substitutability is inherently *reductionist:* it ignores the individuality and specificity of each individual entity by subsuming it under some general value category, thereby making it comparable to others. This answer is, however, as unclear as it is popular. To criticize some way of thinking as "reductionist" implies that one or more aspect of a thing is neglected that is held to be particularly important, thus leading to false or at least distorted ways of thinking about it. An axiological approach to natural entities, however, is not reductionist in any of these ways. It is true, axiology implies comparability, especially with nonlexicographic orderings of values, but this does not imply that the specificity of each single constellation of values present in a natural entity is thereby somehow eliminated. Even if the Big Thicket of Texas and the Palo Verde Canyon get a certain ranking, or even a point value, in the valuation scheme of some nature protection agency, this abstract aggregate value is the result of a number of more specific and more fine-grained valuations taking into account a plurality of value aspects. (Think again of the complexity of factors going into the judgment of the overall "justice" of a certain distribution scheme, for example in the allocation of scarce organ transplants.)

Karl Marx, in his critique of money in *The Capital*, famously called money the "radical Leveller of all differences, . . ." in which "all qualitative differences of commodities are wiped out," thus drawing an analogy to Hegel's concept of the concept in which all particular differences are "*aufgehoben*" (Marx 1965, 145f.). A certain sum of money does no longer show "what has been transformed into it." Likewise, the resultant number of points in a priority list of natural sites worthy of preservation does not show what kinds of concrete values have gone into it. But that amounts to a *reduction* only if not all relevant value aspects have been taken into account. In this respect, the comparison with monetary value is unfair. While the market value of a thing does not in general give an adequate picture of the value ascribed to it (the consumer usually pays less for it than he would be prepared to pay on the basis of his preferences), this need not be the case with a more elaborate and adequate scheme of valuation. A systematic axiological approach to ecological choices is not at all subject to the same criticism as the attitude of the snob who, according to Oscar Wilde, knows the price of everything and the value of nothing.

A second conjecture is that what is at stake is not commensurateness but the fact that by subjecting natural entities to any valuation scheme whatsoever, we are exercising *control* over it. We impose, as it were, our own valuation scheme on something to which human valuation is alien, subjecting nature to some kind of *heteronomy*. The psychological basis of this feeling seems to be this: man has arrived late on the earth. It is not adequate that man as a late product of evolution arrogates to himself the role to decide what in nature should exist and what not. It is simply arrogant to judge what we have not created and could not have created.

But this idea does not lead us very far. Talk of heteronomy makes sense only where talk of autonomy makes sense. Since most parts of nature are incapable of valuations of their own, the terminology of heteronomy is just out of place. There would be "arrogance," to take up Ehrenfeld's title *The Arrogance of Humanism*, in valuing nature and thereby subjecting it to human standards only if we could confidently say that nature has a will of its own.

If there is no "arrogance" in human valuation as such, isn't there "arrogance" in the concrete *act* of changing nature according to these valuations, that is, in the act of actively substituting given natural entities by others, thus changing, modelling, and re-modelling biotopes, ecosystems, and landscapes?

In response to this it may be asked why there is a problem in actively changing nature according to human values when nature itself is a process in constant flux, involving the replacement of individuals, biotopes, ecosystems, and species, though at a much slower pace? What is particularly problematic in anthropogenic changes in contrast to spontaneous changes? Katz in his article of 1985 obviously thinks substitution problematic only if effected by human intervention, not if effected spontaneously, for example, by one biological species driving another out of the system and taking its place. But what, it might be asked, is special about anthropogenic substitution? It cannot be the case that, as Katz seems to imply, in the case of anthropogenic substitution some kind of intrinsic value is lost that is not lost in the case of spontaneous substitution. In both cases, intrinsic value is lost (or gained), depending on what is substituted for what. Apart from possible side effects, the axiological overall value of a natural system is not affected by whether substitution comes about spontaneously or by human intervention. If the intrinsic value of each natural item is bound up with its "integrity and identity" (Katz 1985, 255), this intrinsic value is also lost by natural extinction.

The reason Katz gives for this asymmetry is that only humans are moral agents and are therefore responsible for the loss of intrinsic value involved in substitution. But this seems to misplace the asymmetry. Humans are responsible not only for active interventions but also for forbearances, and often, humans are able to alter the course of events in such a way that a spontaneous process of substitution is prevented. For Katz, "natural" substitutions are never problematic, preventable or not. If weeds grow in my garden and drive out my flowers though I could prevent them from doing so this would, according to Katz, constitute no comparable problem.

What is seen as problematic about substitution, then, is *neither* whether the sum total of intrinsic value is diminished, preserved, or increased in the process, *nor* whether it falls within the range of human responsibility, but whether it involves any *active destruction* of given natural entities. The distinction between *activity* and *passivity, agency* and *nonagency* and not that between these other factors seems to be the central variable.

The theoretical problem with this variable, however, is that it has no place in a purely axiological approach, at least not in axiologies of the nonmoral sort we have hitherto considered. The distinction between active and passive is a characteristically *deontological* distinction. This explains the difficulties to integrate the values of "integrity and identity" into a purely axiological scheme. Seen from a purely axiological point of view, integrity and identity are values, if they are values, of a general sort, not tied to any particular individual. If e_1 are replaced by e_2, the integrity and identity of e_1 are replaced by the integrity and identity of e_2, just as other value properties of e_1, say beauty, complexity, diversity, or productivity, are replaced by those of e_2. The fact that the substitution is effected by the *infringement* both of the identity and the integrity of e_1 has no effect on the value balance.

What underlies Katz's concern about substitution seems, then, a deontological principle prohibiting not so much substitution as such but the acts of active harming natural entities involved in many of them. As an action-related rather than an end-state principle, this principle resists integration into a purely axiological approach. Nevertheless, there remain two problems with Katz's way of stating his position, a meta-ethical one and a normative one.

The meta-ethical problem is that Katz's formulation tends to make what is at bottom a deontological norm of human behavior appear as some kind of intrinsic value of the things which are the object of this behavior. The wrongness of actively destroying natural entities is expressed in axiological language as the value of the integrity of these natural entities. What is misleading about this is that the special logical role of a value of integrity as opposed to other values is left implicit. The same structure can be found in Kant's concept of human dignity, which in fact has led to analogous misunderstandings. Human dignity is a deontological norm that prohibits treating a person only as a means and not also as an end in itself. In Kant's theory, however, this norm is given the logical status of a value of the object of the forbidden instrumentalization. Human dignity is treated as a *property* of human beings. This way of putting it, however, encourages the question why it should be contrary to human dignity to substitute one human being for another (for example by selective abortion and subsequent birth of another child) since the sum total of human dignity is kept constant in the process.

How plausible is a deontological constraint on human substitution of natural entities? Stated as a principle rigidly prohibiting any destruction of natural items, this principle seems hardly acceptable even on a narrow interpretation of "natural items" excluding natural entities that owe their very existence to human intervention. Nevertheless, a principle of this kind *does* seem to underlie Ehrenfeld's and Katz's reservations about the irreversible elimination of variola, the smallpox virus. Destruction of a pest is held by these authors to be a problem even if keeping it alive implies risks of destruction on an enormously greater scale.

A more plausible strategy would be to soften the rigidity of an appropriate deontological constraint by integrating it into a mixed deontological-axiological value scheme as *process-values* that contribute to the overall value of a state of nature without dominating it. An *extended* axiological framework on these lines seems more suited to incorporate the specific negative valuation put by many environmentalists on processes of substitution involving destructive human interventions. At the same time it would not prejudge the issue of what exact weight to give to these values.

An example of such an extended axiological framework is provided by Paul W. Taylor's "ethics of respect for nature" (Taylor 1986). Though Taylor's nature ethics starts from an axiological postulate, the moral rules constructed on this basis are of an explicitly deontological kind. The only intrinsic value in nature recognized by Taylor's axiology is the teleonomic structure of biological organisms. This by itself leaves ample room for substitutions between individual organisms since the value of the teleonomic structure of individuals is assumed to be the same for "higher" and "lower" organisms. Substitution is restricted, however, by a rule of nonmaleficence as well as by a "principle of minimum wrong" (following Regan 1983, 297–312). Both rules prohibit harming a biological organism even if the net sum of the underlying value is thereby increased. The beings that are benefited or harmed by an act of substitution are not regarded "as so many 'containers' of intrinsic value or disvalue" but as beings "to which are owed prima facie duties" (Taylor 1986, 172, 283f.). As with Katz – and differently from Attfield (cf. Attfield 1983, 156) – "harm," in this context, means *active* harm, not, for example, negligence or the failure to prevent harm coming from other sources (Taylor 1986, 172). In contrast to Katz, however, the rule of not harming is stated by Taylor as a prima facie rule which allows for trade-offs with other deontological constraints as well as with the values specific to the human sphere ("respect for persons"). In this way, it accords much better with widely shared intuitions.

9.4 LIMITS TO SUBSTITUTION IN AN AXIOLOGICAL FRAMEWORK: HISTORICAL VALUES

Are there limits to substitution even in an unextended axiological framework? It might be thought that substitution can be satisfactorily excluded by giving some items of an ontology *infinite* value. But this strategy does not preclude substitution provided that what is substituted is also of infinite value. What we have to look for in order to *limit* substitution is a value property that necessarily vanishes, or deteriorates, with substitution, like the genuineness of the original of a work of art vanishes with substitution by a perfect copy. The word "necessary" is important here, since we want to abstract from all *practical* difficulties confronting actual substitutions. From a practical point of view, Nicholas Rescher is certainly right in pointing to the fact that the preservation of endangered species is a much safer strategy to maintain biodiversity than the substitution of lost species by new ones synthesized by genetic recombination (cf. Rescher 1980, 90ff.). But what is the *theoretical* position underlying this warning? No doubt, it is the position (which will presumably shock environmentalists) that overall value would be *preserved* by such a substitution, given that the overall level of existence and the level of diversity are maintained.

The most obvious candidates for axiological values by which substitution is severely limited are *historical* values such as *age* and the fact that something is of *natural* origin (cf. Elliot 1982; Attfield 1994). Age, be it the ontogenetic age of an individual plant, the phylogenetic age of a biological species, or the geological age of a landscape, is an important criterion for quite a number of environmentalists. For example, Eric Ashby (1980, 28f.) writes that it "would . . . be vandalism, and therefore immoral to destroy unnecessarily something which we cannot create and which is the expression (and not the end-product) of millennia of evolution." Though substitution of younger entities by older ones is conceivable, substitution in the reverse direction tends to be the rule, especially where natural systems are harvested for human use, including the slaughter of domestic animals. *Naturalness* and *wildness* understood in a genetic sense are historical values of equal importance, at least where substantial areas of wilderness persist. In Central Europe, where even so-called "virgin forests" are in general no older than a few centuries, efforts at conservation tend to concentrate on naturalness and wildness in a purely *phenomenal* sense, referring not to the *history* of natural entities but to their *appearance*.

As far as historical interest is directed at singular objects with a certain genetic or causal role, objects of historical value like manuscripts, documents, and relics are strictly irreplaceable. Historical values, however, cover only a small part of what we value in nature. Historical values like age and authenticity are relevant to natural monuments (a concept which was coined, not surprisingly, in the period of historicism), but it may be doubted whether they carry much weight, for example, with biological species. Is the spider, as Albert Schweitzer once suggested to one of his guests in Lambarene who was about to kill the insect in his room, a worthy object of protection because it is, as a species, so much *older* than man? Historical value has its legitimate place in a museum, but it is doubtful if the museum perspective should dominate our practical dealings with nature.

Moreover, historical authenticity is a value of only very limited range. We are interested in authenticity where it goes together with additional values, such as causal importance and artistic quality, or with relational values like the role a work of literature has had in one's personal biography (even when it has long lost its personal value as a work of art). It does not seem to make much of a difference how old a component of nature is if it is devoid of all further qualities that recommend it for preservation, or which, like the smallpox virus, strongly recommend it for elimination. Even the protection of biodiversity is a reasonable end only if diversity is combined with other qualities. One may well doubt if a greater diversity of flu viruses is to be preferred to a lesser one (cf. Marggraf and Streb 1997, 238).

We have so far only mentioned in passing the probably strongest reservations against substitution: *personal* values rooted in affective ties like love and friendship. Personal ties transcend axiological value because their objects are not general *properties* but individual *objects*. The objects of love and friendship are inexchangeable precisely because these emotions are directed at individuals *qua* individuals and not *qua* possessors of general properties. That is why love, more clearly than friendship, is an unsuitable object for *why*-questions.

The absence of substitutability in affective relations is, however, strictly *relative*. An object of love or friendship is nonsubstitutable not in the impersonal sense of a philosophical axiology but only with respect to the individual that is affectively related to it. A part of nature may be greatly important for an individual, the population of a certain geographical area, or a whole generation, and have no or very little importance for others living at different places and times. This shows that relational value is ultimately a variant of *extrinsic* and not of *intrinsic* value. This value essentially depends on the relations humans build up toward their natural environment. If there is axiological value in these relations (there certainly is), it must be sought not in the value that is attributed to the objects of these relations but in their contribution to these intrinsically valuable relations, either as causes and causal factors (instrumental value) or as intentional objects (inherent value).

As far as natural entities are concerned, these relations are highly diverse in quality, intensity, and extent. The intentional objects of love, awe, admiration, and sentimental attachment range from single animals, plants, or mountains to the whole of nature, creating, in every single case, a certain resistance against change and substitution. We cling to what we know and have grown familiar with, an effect studied under the names of *endowment effect* (Thaler), *status quo bias* (Samuelson/Zeckhauser), or *loss aversion* (Kahneman/Tversky). People in general demand much more to give up an object than they would be willing to pay to acquire it (cf. Kahneman et al. 1991). The status quo is looked upon as a kind of *possession*, with a corresponding reluctance to replace what already exists with what will exist in the future even if it is certain that the substitute will be a perfect equivalent.

Dieter Birnbacher

9.6 SUBSTITUTION WITHIN THE CONSTRAINTS OF BIODIVERSITY PROTECTION

By way of conclusion, it may be said that it is the very extent of the *variability* of what is loved and liked in nature between cultures and historical periods that is one of the strongest reasons for the protection of biodiversity. It is a strong reason for the protection of diversity in nature in a *conservative* sense, but at the same time it is a strong reason for an active *furtherance* of the evolution of diversity wherever this is possible without jeopardizing other important ecological and nonecological goals. Within the constraints of securing a high level of biodiversity as an ecological and potentially economic safety margin for future generations, however, anthropogenic substitutions of natural individuals and natural systems seem unobjectionable. Substitution may even be ecologically productive, as by counteracting tendencies to ecological simplification, actively supporting endangered species, or replacing common varieties by rare varieties and frequent ones by endangered ones. Nature is not sacrosanct. We should feel free to improve it. Biodiversity might be, after all, a much less conservative value than it appears at first sight.

BIBLIOGRAPHY

Ashby, E. 1980. The Search for an Environmental Ethic. In *The Tanner Lectures on Human Values. Vol. I.* S. M. McMurrin, ed. Salt Lake City/Cambridge: University of Utah Press.

Attfield, R. 1983. *The Ethics of Environmental Concern.* Oxford: Blackwell.

    1994. Rehabilitating Nature and Making Nature Habitable. In *Philosophy and the Natural Environment.* R. Attfield and A. Belsey, eds. Cambridge: Cambridge University Press.

Chang, R. 1997. Introduction. In *Incommensurability, Incomparability, and Practical Reason.* R. Chang, ed. Cambridge, Mass.: Harvard University Press.

Ehrenfeld, D. 1978. *The Arrogance of Humanism.* New York: Oxford University Press.

Elliot, R. 1982. Faking Nature. *Inquiry* 25:81–93.

Frankena, W. K. 1979. Ethics and the Environment. In *Ethics and Problems of the 21st Century.* K. E. Goodpaster and K. M. Sayre, eds. Notre Dame, Ind.: University of Notre Dame Press.

Jakobson, K. M. and Dragun, A. K. 1996. *Contingent Valuation and Endangered Species. Methodological Issues and Applications.* Cheltenham, UK and Brookfields, USA: Edward Elgar.

Kahneman, D., Knetsch, J. L., and Thaler, R. H. 1991. Endowment Effect, Loss Aversion and Status Quo Bias. *Journal of Economic Perspectives* 5:193–206.

Kattmann, U. 1997. Der Mensch in der Natur: Die Doppelnatur des Menschen als Schlüssel für Tier- und Umweltethik. *Ethik und Sozialwissenschaft* 8:123–31.

Katz, E. 1985. Organism, Community, and the Substitution Problem. *Environmental Ethics* 7:241–56. Reprinted in E. Katz, *Nature as Subject. Human Obligation and Natural Community*. Lanham, Md.: Rowman and Littlefield, 1997.

Leopold, A. 1949. The Land Ethic. In A. Leopold, *A Sand County Almanac and Sketches Here and There*. New York: Oxford University Press.

Marggraf, R. and Streb, S. 1997. *Ökonomische Bewertung der natürlichen Umwelt. Theorie, politische Bedeutung, ethische Diskussion*. Heidelberg: Spektrum.

Marx, K. 1965. *Das Kapital*. vol. 1. Berlin: Dietz.

Regan, T. 1983. *The Case for Animal Rights*. London: Routledge.

Rescher, N. 1980. Why Preserve Endangered Species? In N. Rescher, *Unpopular Essays on Technological Progress*. Pittsburgh, Pa.: University of Pittsburgh Press.

Taylor, P. W. 1986. *Respect for Nature. A Theory of Environmental Ethics*. Princeton, N.J.: Princeton University Press.

Wilson, E. O. 1995. Jede Art ein Meisterwerk. *DIE ZEIT* 23. 6. 1995, 33.

Part IV

Protecting Biodiversity

10

Biological Diversity and Conservation Policy

KATE RAWLES

This chapter is about biological diversity, or biodiversity – one of those things everyone is in favor of. A rough definition of biological diversity is *the diversity of living things at a number of levels: genetic, between and within species, and between and within habitats and ecosystems.* The rest of this chapter could be spent exploring the adequacy or otherwise of this definition and the conceptual issues that arise in relation to it. I am not going to do that. Instead, I want to raise three questions about the role of biological diversity in the formation of conservation policy.

The first question arises in the context of conflict between conservationists and animal welfarists over issues like culling. The question is, *does the aim of preserving biological diversity justify conservation policies that involve killing sentient animals?*

Conservation aims are increasingly formulated in terms of the preservation of biological diversity. My second question is, *does the aim of preserving biological diversity fully capture what conservationists actually do?* I suggest that it does not. There is an aspect of conservation irreducibly concerned with the dubious concept of nativeness.

Third, I want to ask *whether the preservation of biological diversity is what conservationists should be trying to achieve.*

10.1 DOES THE AIM OF PRESERVING BIOLOGICAL DIVERSITY JUSTIFY CONSERVATION POLICIES THAT INVOLVE KILLING SENTIENT ANIMALS?

Bryan Norton (1987, 156) writes that "[c]oncern for biological diversity stands, in a sense, as the most central value of environmentalism, because other environmental goals such as resource protection, pollution abatement and so forth all depend on the continued functioning of complex ecosystems." In the last ten years or so, biological diversity has also become a central and guiding principle of conservation, largely displacing other candidates such as ecological integrity, wilderness, wildness, or naturalness.

The aim of preserving biological diversity supports the tendency within conservation to value and to focus attention on entities such as species, populations, habitats, or ecosystems. Conservationists – I am referring to "nature" conservationists rather than those concerned with, say, historic buildings – do not, as a rule, value or take much interest in *individual* living things, other than in a derivative or indirect sense. For example, conservation resources may be invested in protecting the nests of very rare birds, such as ospreys and peregrine falcons, but only as a means of conserving the species. Were these birds to become more common, the protection would cease. The overall aim of preserving biological diversity reinforces this primary concern with species and other ecological collectives, rather than with individual animals or plants.

The animal welfare movement is quite different.[1] In contrast to conservation, the animal welfare movement is primarily concerned with individuals rather than collectives, and with individuals of a certain kind: those that are conscious and able to experience pain and/or pleasure. I will follow Peter Singer (1991, 8) in referring to these as "sentient" individuals. The overall aim of the animal welfare movement is to minimize or prevent the human-caused suffering and death of sentient individuals – commonly taken to be a subset of animals. Habitat protection may of course be necessary to achieve this. But the concern for habitats, ecosystems, or other nonsentient entities is secondary or derivative. Animal welfarists do not normally acknowledge any direct ethical obligations to ecological collectives.

These essentially philosophical differences between the two types of organization lead to very different practices and policies in the field. Sometimes these are complementary. Consider, for example, reintroduction and rehabilitation. Conservation bodies may attempt to re-introduce species to an area where they were previously found, particularly if that species is rare or endangered. Sea-eagles were reintroduced to the Scottish Island of Rum using birds flown from Norway in Royal Air Force fighter jets. This kind of project is immensely costly in terms of both time and money, but is considered worthwhile – especially when, as in the Rum case, the reintroduction is successful. Conservation bodies will not, however, take on the rehabilitation of injured individual wild animals. This is considered too costly, and does not contribute to conservation objectives, especially if the animal in question is a common one. Thus, if you turn up at a Scottish conservation site with an injured pigeon, at best you will be told to go elsewhere. Many animal welfare organizations, by contrast, have extensive wildlife centers, and will accept and attempt to treat and rehabilitate any injured animal or bird. But animal welfare organizations do not get involved with species reintroduction.

This amounts to a fairly sensible division of labor. But there are also cases where conservationists kill sentient animals, as a routine part of their management strategies. This has led to considerable conflict between animal welfare groups on the one hand and conservationists on the other, not just in Britain, but around the world.

10.1.1   Ruddy Ducks

Perhaps the most notorious example is that of the ruddy duck. The ruddy duck was introduced to Britain from North America in the 1940s by, ironically enough, the naturalist Sir Peter Scott. It escaped from his wildfowl sanctuary in 1952, and there are now almost four thousand in Britain. In winter, they fly off to Spain, where they mate with the much rarer white-headed duck. The offspring do not have white heads, and the fear is that the white-headed duck will be lost altogether. At the request of the Spanish government and various conservation organizations, the British government has given the go-ahead for thousands of these ducks to be culled. The cull is to be achieved by shooting female ducks, as they sit on their nests. The cull has been approved by the Royal Society for the Protection of Birds, a conservation organization. It is strongly opposed by animal welfare groups such as Animal Aid and the League against Cruel Sports.

The ruddy duck is a particularly bizarre case. Different species do not, as a rule, interbreed under normal circumstances; and the ruddy and white-headed ducks are not in fact different species. This raises interesting questions about the significance of subspecies to the preservation of biological diversity. But, however these questions are answered, the ruddy duck case is by no means the only example of routine conservation policy that involves killing sentient animals.

New Zealand, for example, has no indigenous mammals other than a single bat, and has evolved a unique population of "dopey" birds which are extremely vulnerable to imported mammals such as cats and foxes. Conservationists are regularly required to kill cats and foxes in New Zealand, and also in Australia. Elephants are killed in Zimbabwe because of their alleged impact on the diversity of the national parks. In Scotland, there is an annual cull of red deer, whose large numbers prevent tree regeneration. In the English Lake District, grey squirrels are killed in the attempt to prevent the loss of the native red squirrel. Introduced escapee mink are hunted the length and breadth of Britain. In all these cases, conservationists and animal welfarists are at loggerheads.

The primary focus on species and other ecological collectives on the one hand, and on sentient individuals on the other, partially explains this conflict. Further light can be shed on it by considering other differences between the two kinds of organization.

The conservation movement draws on science, and particularly ecology and conservation biology, for authority, validation, and direction. While not advocating avoidable suffering or cruelty, it may dismiss what it sees as excessive concern for individual animals as sentimental. Conservationists often view the animal welfare movement as overly emotional, irrational and, in the end, misguided. From this perspective, to focus on individual animals is to miss the wider and more urgent picture of species extinction and biodiversity loss. Concern with biological diversity does not seem to entail that suffering should be reduced. Indeed, conservationists may argue that suffering, however caused, is an inevitable part of life, and should be accepted as such. On the other hand, introduced species are held to be the second biggest threat to biological diversity worldwide, and culling is therefore simply a necessity, if biological diversity is to be preserved.

The animal welfare movement draws on philosophy and particularly ethics for its authority and validation. It can perceive conservationists as excessively managerial, God-playing, ruthless, and generally in the grip of a heartless science – and, ultimately, unethical. The first biggest threat to biological diversity is human-caused habitat extinction. But we don't advocate culling humans.

10.1.2   Animal Welfare Arguments

This alleged double standard is the nub of the animal welfare argument, or one version of it. These arguments are well-rehearsed and I will only summarize them briefly here. The first starts from the assumption that we have ethical obligations to individual humans. It then argues that there are no morally relevant differences between humans and other sentient animals. On the other hand, there are morally relevant similarities, such as the capacity to suffer. Thus, simple consistency requires us either to deny our obligations to humans or accept that we have them to other animals too. To do otherwise is to be guilty of speciesism; the discrimination against a creature purely by reference to its species, a characteristic held to be as irrelevant, in this context, as race or sex.

The second argument focuses on the relevance of characteristics such as sentience. This argument comes in two stages. The first part involves a view of what it is to act morally. For example, it may be argued that to act morally is to consider the interests of others, and not just one's own; or that the essence of morality has to do with restraining from causing harm to others. Second is the claim that consciousness and the ability to suffer – sentience – are the capacities that make it possible for a creature to have interests, or to suffer harm. Thus sentience becomes, in effect, a criterion of moral significance, of being the kind of entity toward which a moral agent can have moral obligations. That animals share this characteristic with humans is not crucial to arguments of this kind.

Both types of argument conclude that we have clear ethical obligations to individual sentient animals, and that these obligations involve a concern for their lives and well-being.

10.1.3   Commentary

I want to make three points here: two about the animal welfare argument, and one general one about culling.

First, let me point to the irony inherent in the accusation of emotional over-sentimentality leveled at animal welfare campaigners by conservationists, among others. In fact, the arguments that animal welfarists draw on, articulated in the philosophical literature by, for example, Peter Singer (1991), Tom Regan (1984), and Bernie Rollin (1992), explicitly *disavow* any appeal to emotion, utilizing instead a very hard-nosed appeal to consistency and logical reasoning. In my view, this approach is if anything *too* rational, leaving no room for the legitimate role of emotions in ethical deliberation and underpinned by a mistaken view of what emotions are like. Very many people, on learning that nesting ducks are shot at their most vulnerable, or that the drays of grey squirrels are smashed and their kittens stamped on, and that these actions are performed, not by vandals but *conservationists*, experience a range of emotions, often powerful ones. Acknowledging and reflecting on these emotions seems to me to constitute a key part of developing an ethical response to such actions. But this is the subject for another paper.

Despite reservations about the neglect of emotions in these kinds of arguments, the animal welfare position overall is in fact a strong one. This is my second point. In my view, the claim that we have ethical obligations to conscious, sentient animals, on the basis of their capacity to suffer, is hard to refute. These obligations are additional to any we have to nonsentient life-forms. If a creature can feel pain, then our treatment of it raises ethical issues that our treatment of nonsentient beings simply does not. Thus, culling squirrels has a further ethical dimension to it than culling rhododendrons.

On the other hand, it is not actually at all clear that we have ethical obligations to entities such as species, habitats, and populations; nor to biological diversity per se. Such entities are not conscious and cannot suffer. Arguably, they have no interests and cannot be harmed. Of course, given that the lives and welfare of all sentient beings can be said to depend on the preservation of biological diversity in various ways, we clearly have very significant indirect ethical obligations here, if not direct ones. As an aside, let me say that I do, in fact, believe that some sense can and should be made of the idea of direct ethical obligations to a wide range of nonhuman natural entities. But I also think that these obligations are extremely hard to articulate, particularly within what might loosely be called a "rationalist" ethical framework, that is, one which takes itself to appeal only to reason and which eschews appeal to emotion. Arguably, it is a re-evaluation of these kinds of frameworks that is called for here, rather than a limiting of our ethical obligations.

My third point concerns culling. Even if it could be shown that killing sentient animals *could* be justified in the pursuit of conservation objectives, the claim that such culling is necessary in order to preserve biological diversity often oversimplifies a complex context in which economic or other human interests are playing a crucial part, and in which the targeted animals are effectively scape-goated.

For example, red squirrels favor coniferous woodlands, whereas grey squirrels thrive in mixed or deciduous woodlands. Long-term, a policy of planting and managing trees for the red squirrel would, arguably, be more effective than killing grey squirrels (see for example Laidler 1980). But it would be much more costly as well. And planting coniferous forests often conflicts with other conservation objectives, such as the regeneration of native oak woodlands. Hence it is said that we "have to" kill the greys to save the reds – and this is done, even though many conservationists also believe that this strategy will ultimately be unsuccessful. Mink released in Britain are said to have adversely affected local wildlife. But, where the otter population is healthy, otters readily out-compete mink and effectively control their numbers. Otter populations, however, have been decimated by anthropogenic river pollution. Again, it is usually cheaper to kill mink than to clean up and remove the source of river pollutants. And with the ruddy ducks, my understanding of the situation is that the population of white-tailed ducks had already been severely affected by Spanish shooting practices – only recently constrained by conservation concerns – and by human-caused pressure on its habitat. In Spain, the white-tailed duck now survives only on Lake Cordoba, a vastly reduced range.

To expand this range again would be a far more effective way of improving the prospects for white-headed ducks. It would also be harder, costlier, and less politically expedient than lobbying for a cull of ducks elsewhere.

In sum, my first question was, does the aim of preserving biological diversity justify conservation policies which kill sentient animals?

I am not at all sure that it does. It is clear that we have direct ethical obligations to sentient animals; but it is not at all clear that we have direct ethical obligations to entities such as species, or to biological diversity. The burden of proof should thus be on conservationists to show how killing the first to preserve the second can possibly be acceptable from an ethical point of view.

10.2   DOES THE AIM OF PRESERVING BIOLOGICAL DIVERSITY FULLY CAPTURE WHAT CONSERVATIONISTS ACTUALLY DO?

Conservation policy is increasingly formulated in terms of the preservation of biological diversity. But biological diversity does not fully capture what practicing conservationists are really after. There is an element in conservation goals which is irreducibly to do with preserving the native, or the natural or, perhaps, with preserving a historical lineage. To illustrate this, I will consider the culling, not of animals, but of plants.

10.2.1   Rhododendrons

Consider, for example, the shrub Scottish conservationists – and those throughout Britain – love to hate: *rhododendron ponticum*. Rhododendron was imported to Britain from Spain and Portugal in 1763. The ongoing attempt to eliminate or at least control this spectacular plant consumes vast amounts of time and money (and is at best only partially successful). Gorse, on the other hand, a vivid yellow plant beloved in its native Scotland, is detested in New Zealand, where similar efforts are made to uproot it.

The reason given for the attempts to eliminate rhododendron from Scotland and gorse in New Zealand is that they are not native. What is so objectionable about non-native plants? A common answer refers to their tendency to outcompete native species, and/or to provide at best a greatly impoverished habitat for them. Both tendencies are held to arise from the plants having been transported into an ecosystem they have not evolved within. In the case of rhododendrons, for example, British animals and insects find it inedible; its roots produce herbicidal chemicals that suppress the growth of

205

other vegetation nearby; and it becomes so dense that nothing can live beneath it. Similar things are said of sycamore, Japanese knotweed, giant hogweed, and Himalayan balsam. The introduction of non-native plants is thus held to result in a loss of biological diversity in that particular area. Hence, for conservationists seeking to preserve biological diversity, while native species are valued, non-natives are not.

10.2.2 Species Diversity

What has been said so far suggests that the bottom line for conservationists is the preservation of biological diversity; and that native species are valued as a means to achieving this, rather than for independent reasons. However, this suggestion begins to look less convincing if a bit of analytical pressure is put on the notion of biological diversity and what is meant by preserving it.

We can start by drawing a distinction between biological diversity at the level of species and at the level of habitat, and by asking which level of diversity it is that conservation seeks to conserve. I will begin with species. Is concern with species diversity the underlying, guiding goal of conservation? Various considerations suggest that it is not.

First, suppose it turned out that an introduced species *did* support as many insects and so on as the native plants it displaced, *and* that it was more rigorous. If species diversity were the guiding goal of conservation, then conservationists should welcome such a plant. But, the majority of conservationists with whom I have discussed this insist that even if sycamores, say, began to support a massive local insect population, they would still not be welcomed in British conservation sites, because, well, because they are not British.

Second, if species diversity were the underlying goal of conservation, then, given the possibility of *adding* some species to an area that was naturally short of them, but that could support a few more, conservationists would jump at the chance. Britain, apparently, is just such a place, because it was cut off from the rest of the European landmass before what could have been its full complement of species had a chance to get there. But the idea of adding some French flora and fauna, say, to Britain, even supposing that this would result in more rather than less species diversity, and that this could be sustained over time, typically fills conservationists with horror.

10.2.3 Habitat Diversity

Perhaps, then, it is diversity at the habitat level that conservationists are really after.

Preservation of habitat diversity would involve the preservation of habitats that are considered unique. Having a lower number of species than it could have is what is unique about Britain. Thus, concern with habitat diversity in this extended sense would explain why conservationists don't favor filling Britain up with imported species.

Similarly, the preservation of habitat diversity would require the preservation of habitats low in diversity as well as those teeming with species. This would explain why conservationists often go to considerable lengths to conserve habitats such as chalk grasslands, which are preserved and prevented from turning into scrub and, eventually, oak woodland, even though the woodland would offer greater species diversity.

A related point concerns the rationale behind the allocation of conservation resources. If species diversity were the ultimate goal of conservation, all conservation resources should, as several commentators have pointed out, be invested in Indonesia and nowhere else. But if it is habitat diversity that conservationists are really after, then they can justifiably mess about conserving bits of Europe and other places where whole hectares of land have lower levels of species diversity than a single Indonesian rainforest tree.

Again, this seems to suggest that native habitats, and the species that constitute them, are conserved because this is the best way to conserve diversity of habitats overall, in which case, nativeness is again being valued only as a means to an end.

But if we explore some of the implications of this conclusion it begins to look much less plausible. Suppose that habitat diversity across the world turned out to be preserved or even enhanced by encouraging the development of patches of rainforest in Scotland, desert in the south of England, and Atlantic heath in Indonesia. Conservationists would surely reply that this is not what they want. Re-introducing eagles to Rum is one thing, because they were there before. But *introducing* a habitat (even supposing this to be feasible) would be quite another. *Even if overall habitat diversity were enhanced*, such a project would not be favored because it would involve introducing something with no historical precedent.

A similar point can be made at the species level. Suppose that global diversity turns out to be preserved or even enhanced by letting introduced species do their own thing. On this scenario, biological diversity would, if we let it, simply be, as it were, redistributed; rhododendrons would settle in Wales and gorse in New Zealand and the ecological systems would in time reform and prosper around them. Most conservationists would strongly favor the preservation of the biological diversity endemic to particular places – even if allowing the non-natives to settle in would ultimately result in as much diversity worldwide. And this suggests that nativeness is in fact valued in its own right.

10.2.4   Nativeness

Now, I am no fan of the concept of nativeness. I would argue, in fact, that the distinction between native and introduced species rests on criteria of dubious merit and inevitably includes an arbitrary element. Moreover, I am sympathetic to the claims made, for example, by Judy Ling Wong of the Black Environmental Network, that it has racist undertones, especially when the debate about plants is simplified to native good, non-native bad (Wong 1999).

Of course, none of this is to deny that rhododendrons in Britain do not support much insect, plant, or bird life. But the same is said of bracken, an indigenous plant. And the opposite is said of Buddlea, the so-called butterfly plant which we are all exhorted to grow in our gardens, and which Judy Ling Wong (ibid.) describes as "a Chinese plant with a good British passport." The question of a particular plant's impact on its surrounding ecology could and perhaps should be separated from the question of whether it is indigenous. Moreover, we could usefully avoid the language of natives and aliens altogether.

The point I want to stress here, however, is that concern with nativeness, or naturalness, or historical integrity, is present in conservation, as an end in itself. It is a separate value that underpins and guides conservation policy, and one that is not reducible to the value placed on biodiversity preservation. This is interesting, and potentially problematic, in the context of the current political climate in which, as I have said, conservation organizations are increasingly formulating their goals in terms of biodiversity – and in which this is in part because of pressure from national governments to fulfill the objectives of the biodiversity convention and from, for example, the European community to meet objectives set out in the habitat directives.

Perhaps, then, the preservation of biological diversity is what conservation *ought* to be about, even if it isn't already. So, here is my third question:

10.3   SHOULD THE PRESERVATION OF BIOLOGICAL DIVERSITY BE PRESCRIBED AS THE GUIDING AIM OF CONSERVATION?

It is hard to deny that the concept of biological diversity, or biodiversity, has been a useful one for conservation. Unlike alternative "banner" concepts such as wilderness or integrity, biodiversity is specific enough for us to get a handle on it, without being so specific that a wide range of concerns can't be rallied beneath it.[2]

That said, when the preservation of biological diversity is advocated as *the* guiding principle for conservation policy, many reservations have been expressed about it. I will mention four in passing and focus on a further two that perhaps concern me most.

10.3.1   Reservations about Biological Diversity

It has been suggested, by a range of commentators, that the emphasis on the preservation of biological diversity:

(i) Encourages a focus on species, and the management of sites for species, rather than on ecological processes;

(ii) Is an inherently static concept that does not encourage acknowledgment of the constant evolution of natural systems;

(iii) Is underpinned by the "natural balance" paradigm; a model of ecological systems that is increasingly rejected in favor of the view that ecological systems are chaotic and in constant flux (Perretti 1998);

(iv) Is inherently managerial. Biological diversity is never discussed without a verb. Unlike nature, that just is, biological diversity is always being preserved, conserved, maintained, or even enhanced. Hence the concept is implicitly managerial, and biased against a more hands-off approach to conservation which some favor, or at least want to keep open as an option (Evans 1996).

All of these criticisms are contentious. They are not meant to deny the existence of rigorous debate within conservation about how the notion of biological diversity is best understood; and they are aimed at the way in which concern for biological diversity has been translated into policy at national and international levels rather than at conservationists themselves. I mention them to indicate the existence of a range of reservations about biological diversity that are interesting and that would merit further discussion. Here, though, I will move on to two further complaints that I want to linger on in finishing. These are that the concern with biological diversity:

(v) Does not capture what is actually significant to people about the natural world; and hence

(vi) Does not motivate people to care about conservation.

10.3.2   The Transfer of Significance

With regard to (iv), here is an alternative way of characterizing the aims of conservation:

> Conservation is about negotiating the transition from past to future in such a way as to secure the transfer of maximum significance.

This is a characterization of conservation that is usually attributed to myself and Alan Holland (Holland and Rawles 1993). Since it was in fact devised by Alan single-handedly, I feel I can advocate it without undue hubris. Both of us now have reservations about "maximum." What I still like about it is "significance."

So, if conservation is about transferring significance, what is significant? What I like about this definition is precisely that it doesn't answer that question. And this effectively opens it up for discussion and debate.

Having opened such a debate, clearly I am not going to do justice to it in my remaining thousand words. But let me at least make a start with the observation that the significance which conservationists emphasize, whether this is expressed in terms of biological diversity, or a concept such as nativeness, or both, is not universally shared.

Busloads of people arrive in Wales every year to enjoy the spectacle of entire hillsides turned purple and pink with rhododendron flowers. What is significant about them for many people is their beauty, particularly in terms of their visual impact, rather than their relation to a scientifically informed notion of biological diversity. Similarly, with regard to ruddy ducks, people often comment on their "perky" appearance, and, ironically, on the rich color of their heads.

From the conservation perspective, this is often held to be unfortunate. The problem, allegedly, is that many people do not understand the scientific context surrounding the ruddy ducks. Hence they respond to essentially irrelevant characteristics such as color. Moreover, in the context of shooting them, many people are disturbed by the thought that they are shot on their nests, while attempting to bring up their young, an activity we both empathize with and value greatly. But this, the conservationist may well say, is anthropomorphism, and should be rejected.

I take this sort of position to be represented by Kristin Shrader-Frechette, who writes: "If public preferences are not aligned with good ecological thinking, the public needs to be educated so that its aesthetic and anthropocentric preferences...take account of the importance of habitat preservation" (Shrader-Frechette and McCoy 1994, 190). I think this is an (uncharacteristically) objectionable statement, and I want to make four points in response to it.

10.3.3 Four Responses

PREFERENCES AND VALUES. First, I think there is often an underlying worry in this kind of comment that if we consult "the public" about what our conservation objectives should be, we will simply have to accept whatever they say. This seems to be rooted in an assumption that Frechette and others appear to make, namely, that while ecologists have conservation values that are scientifically informed, "the public" can only have subjective preferences in this regard. I take preferences to be entirely personal and largely unthought through things that it would indeed be hard to adjudicate between, or to assess in any critical way. It is the assumption that "the public" only has preferences that leads to decision-making techniques such as contingent valuation, where preferences are identified, allegedly rendered commensurate, and simply added up. By contrast, conservationists who partake in "good ecological thinking" are taken to have something more like values, in the sense of things that emerge as the result of reflection, experience, dialogue, and relevant information. Values, unlike preferences, can be critically assessed.

But why assume that "the public" only has preferences and not values? "The public" can engage in critical reflection about nature and still conclude that the things they find of value there are different from those prescribed by the culture of biological diversity. Conversely, those advocating biological diversity can do so thoughtlessly. So, in advocating attention to what people who do not happen to be conservationists find significant in nature, one can still be advocating attention to values; and not, preferences to values that are open to debate, discussion, and modification. This is far from simply accepting whatever the public happens to prefer. But it may lead to a different conclusion about what is significant in nature from that favored by conservationists.

ANTHROPOMORPHISM. My second point concerns the pejorative use of the word "anthropomorphism." Frechette does not use it above, but conservationists and others often do, and often in the context of critical comment directed toward anyone defending the claims of individual animals.

Anthropomorphism, understood as the attribution of characteristics or features that are possessed by humans, to nonhumans, cannot be an error in itself. If it were, talking about beetle legs or earthworm skin would be mistaken. It is the *inappropriate* attribution of such characteristics or features that must be the real target here. But to assume that an empathetic response to another creature's attempts to rear its young, for example, is based on inappropriate anthropomorphism is to deny either the significance, or, more likely, the existence, of nonhuman social relations and subjective mental states such as fear, protectiveness, distress, and so on. This is misplaced. No one who works with animals, as Vicki Hearne points out so convincingly in *Adam's Task* (1982), would deny the existence of animal emotions; and certainly no one who trains animals could do so and have any success as a trainer.

But if other animals do experience such a range of emotions, denying their significance seems very close to what Singer calls speciesism. Why should human grief and loss matter but not that of other species? In my view, it is precisely this ability of nonhuman animals to experience grief, loss, joy, boredom, and to have social relations that matter to them, that renders our relationships with them essentially ethical, and which makes our frequent neglect of these ethical constraints so significant, and so sad. So a simple rejection of "anthropomorphism" is unacceptable.

VALUES AND SCIENTIFIC AUTHORITY. My third point concerns the assumption that the values allegedly derived from what Frechette calls "good ecological thinking" should take priority over other values such as aesthetic or cultural ones. Such an assumption is implicit in the suggestion that what the public needs is an ecological education. But it calls for a clear defense; and the value judgments involved in making it cannot be eliminated by implicit appeal to scientific authority.

Appeal to scientific authority is, of course, not uncommon in conservation. Nor is the denial of the need to make value judgments. A conservation colleague told me that 90 percent of his colleagues in the British Wildlife Trusts would claim that the Biodiversity Action Plan is a value-free document. Clearly this is simply wrong. Value judgments are involved throughout. Value judgments have to be made, for example, about what level of biodiversity is to be pursued (species, habitat, genetic); about what counts as diversity within levels (is the difference between a blue tit and a coal tit less significant than the difference between a blue tit and a reindeer?); and about how conflicts between these different levels are to be resolved. (The previous Scottish Natural Heritage warden on Rum, for example, told us that while he wanted to manage for species diversity by encouraging regeneration of trees, the European community had instructed him to manage for Atlantic heath, because this was a habitat listed under the European habitats directive and Rum has, or is, a particularly good example of it.)

Similar points can be made about nativeness and naturalness. And of course, conservation rests on the fundamental judgment that biological diversity, or nativeness, or naturalness – however understood – is valuable and worth conserving. None of these value judgments can simply be read off from scientific ecology. Even if they could, it would not be evident why value judgments derived from science should take priority over judgments with other sources.

Thus, my third and perhaps main point is quite simply that *there is a genuine debate to be had about what is valuable, or what is of significance, in nature.* There are also accompanying issues about who should contribute to this debate, what counts as authority and expertise in this context, and so on. This debate is essentially an ethical or at least a normative one; and it should be opened up. It should not be foreclosed by automatically prioritizing allegedly scientific criteria. However, and without wishing to make unfair generalizations, it might be suggested that the conservation movement, in Britain at least, is not always engaged or interested in such a debate, and that it may be too quick to assume that the public needs educating rather than consulting.

The work of Jacquelin Burgess and others is relevant here (see Harrison, Limb, and Burgess 1987). Harrison et al. argue that what many nonconservationists value in nature is regular encounters with nature, and particularly with animals. These animals don't have to be rare, or native, or important for biological diversity but, rather, conspicuous. Burgess talks about the delight people take in commonplace encounters with wildlife such as grey squirrels and urban foxes, and the desire that such encounters continue as a feature of their and their children's ordinary, daily lives. It is at least not obvious why these ways of valuing nature should be considered irrelevant. But they are currently sidelined by mainstream conservation policy and practice, especially in its emphasis on site-based conservation. To make just one point, the British emphasis on Sites of Special Scientific Interest does little or nothing to increase contact with nature for most people. They are selected on quite different grounds and the majority of these sites are simply not accessible to people who live in urban contexts – particularly those who do not have cars.

TACTICS. My fourth point follows from this. It is a tactical one, about gaining support for conservation.

Much of the research conducted in this context concludes that what people value about nature depends on their relationship to it, and what it means to them in their everyday lives. And this depends on the kind of connections they have, or have had, with the nonhuman world, particularly as children.

But it is a commonplace, of course, that many people are increasingly alienated from contact with nature. Finding ways of sustaining connections between people and the nonhuman world will thus be crucial if people are to continue to care about it. People will not vote for the preservation of species if this has no significance for them. Yet for many, the language of biodiversity is, to quote John Pollock (1996, 3), a "discourse draining away personal meaning in favour of corporate blandspeak." Arguably, it exchanges what is valued for what is countable. The public briefing on ruddy ducks, for example, throws its scientific weight around, battering the public with "scientific weighting which gives it an aura of absolutism; Latin names, references, distribution maps, graphs, tables and strategies" (Lawson 1996, 27) and completely refusing to engage with anything remotely like empathetic concern for nesting creatures. But it is precisely the latter sort of concern that motivates conservation, that gives it meaning and significance.

Of course, many conservationists, off-record at least, speak about their work in a very different way. When I interviewed the conservationist in charge of the red squirrel project in Northern England, she began by talking about biodiversity and native species and ended up talking about a passionate sense of loss, a profound love of the nonhuman world, and a desire to compensate for some of the damage she takes us to have already done. This was a practicing conservationist working with a conventional conservation body not known for its radical views. Paul Evans (who is none of these things) is thus not completely off the mark when he writes that

> there are many other significances . . . that conservationists may want to artic-
> ulate, other than diversity: wildness, chaos, change, "otherness" . . . many con-
> servationists do have extraordinary insights into what significance in Nature
> actually looks, sounds smells and feels like. Articulating this in a way which
> has a purchase on the language of everyday experience requires skills . . . of per-
> ception and communication. A reliance on science prevents the development
> of these skills. (Evans 1996, 10)

It is the passionate, empathetic response, based on connection with the non-human world, that motivates concern for its conservation. It is this that needs to be nurtured. Reflective empathy and a sense of connection is the life-blood of conservation. But policies articulated in purely scientific terms, and those that prescribe the routine culling of sentient animals, involve shutting down compassion and empathy, and distancing rather than connecting.

10.4   CONCLUSION

Sentient and nonsentient living things are routinely killed by conservation, in the name of biological diversity. I have tried to suggest that killing sentient animals raises significant ethical concerns and that it is at least not obvious that these can justifiably be overridden in the name of biological diversity. This is underlined by the thought that biological diversity does not really capture what many people find significant about nature, which is much more orientated toward particular relationships, meaning, and connection.

This is not to say that we should jettison concern for living things at the species and system levels. But we should find a way of articulating this concern that motivates and inspires, that captures what is held to be of significance in nature, and that does not alienate people from the significance they already acknowledge. And we should find ways of translating this into policy that respects our ethical relationships with other living things, at the level of the individual and not just the species. In sum, we need to put the life back into biological diversity.

NOTES

1.   I use "animal welfare" as an umbrella term to refer to those who recognize direct ethical obligations to individual sentient animals, and who use this to argue for better treatment of these animals by humans. For the purposes of this chapter, I do not wish to distinguish between animal welfare, animal rights, and animal liberation.
2.   Thanks to Greg Mikkelson for this point.

BIBLIOGRAPHY

Evans, P. 1996. Biodiversity, Nature for Nerds? *Ecos* 17(2):7–12.
Harrison, C., M. Limb, and J. Burgess. 1987. Nature in the City – Popular Values for a Living World. *Journal of Environmental Management* 25:347–62.
Hearne, V. 1982. *Adam's Task: Calling Animals by Name*. New York: Alfred A. Knopf.
Holland, A. and K. Rawles. 1993. Values in Conservation. *Ecos* 14(1):14–19.
Laidler, K. 1980. *Squirrels in Britain*. Newton Abbot, North Pomfret: David and Charles.
Lawson, T. 1996. Brent Duck. *Ecos* 17(2):27–35.
Norton, B. G. 1987. *Why Preserve Natural Variety?* Princeton, N.J.: Princeton University Press.
Perretti, J. H. 1998. Nativism and Nature: Rethinking Biological Invasion. *Environmental Values* 7(2):183–92.
Pollock, J. 1996. Negative Science, Positive Public: a Jeremiad on 'Biodiversion.' *Ecos* 17(2).
Regan, T. 1984. *The Case for Animal Rights*. London: Routledge.

Rollin, B. 1992. *Animal Rights and Human Morality*. Rev. ed. Amherst, N.Y.: Prometheus.

Shrader-Frechette, K. and E. D. McCoy. 1994. Biodiversity, Biological Uncertainty and Setting Conservation Priorities. *Biology and Philosophy* 9:167–95.

Singer, P. 1991. *Animal Liberation*. 2nd ed. London: Thorsons.

Wong, J. L. 1999. Presentation at BANC/National Trust Conference, Lancaster University, June.

11

Beavers and Biodiversity

The Ethics of Ecological Restoration

CHRISTIAN GAMBORG AND PETER SANDØE

Ecological restoration has been portrayed recently as a process capable of reversing the loss of natural biodiversity now occurring in many densely populated areas and intensively managed landscapes in Europe (Throop 1997; Hobbs and Norton 1996). Species restoration schemes operate throughout Europe and in parts of North America as well. For example, they have involved the lynx in Poland, and the wolf and the moose in New York State. Human subsistence activities, such as hunting and agriculture, have resulted in losses of wildlife species. Natural environments have been intensively utilized for many centuries, especially in Western Europe and parts of North America; and a high level of productivity characterizes these domesticated environments (Nash 1989). As a result of these efforts to transform the natural environment into a highly efficient growth medium, variation is lacking and natural biodiversity has declined.

Species have died out regionally, and their opportunities to return to former haunts have been seriously limited by intensive management of the natural environment (Thomas 1992). Moreover, artifacts such as roads, towns, and bridges, as well as the straightening of rivers, block the paths of migrating wildlife. New policies on the conservation of wildlife, and on the general management and protection of the natural environment, are pursued in many affluent industrialized countries. These aim to recreate and maintain the dynamics and variation of natural ecosystems (Kane 1994; OECD 1999). This presents new opportunities for the conservation discipline (Pickett and Parker 1994). According to Jordan (1994), ecological restoration may well become just as important as a conservation tool as wilderness preservation.

Restoration is the attempt to reverse human impact by restoring, or returning, an ecosystem or habitat to an earlier state – its so-called "predisturbance

217

situation." In this sense, it has been described as trying to turn back the environmental clock. In other words, restoration attempts to copy a specific historical structure. Certain restoration efforts are perhaps most aptly characterized not as turning back the environmental clock but as "making it tick again" (Cowell 1993). For this reason restoration has been viewed as a variety of "creative conservation" (Sheail et al. 1997). Standard examples of restoration practice include the elimination of introduced (i.e., technically exotic) animal or plant species, the reintroduction of formerly native species, and the large-scale alteration of entire landscapes.

However, while it is generally recognized that biodiversity has been lost, and continues to be lost (Tilman 2000), and while it is widely acknowledged that steps must be taken to resolve this problem, experts disagree over whether ecological restoration in general, and more specifically reintroduction, are effective remedies. The issues raised by the use of restoration ecology to protect biodiversity cannot be settled solely on the basis of prudential considerations. We argue that disagreements pertaining to species reintroduction which superficially appear to be about "factual" biological and managerial issues really stem from fundamentally different conceptions of the value of nature in general and biodiversity in particular.

In this chapter we will use the case of beaver reintroduction in southern Scandinavia to illuminate the philosophical issues underlying the value of biodiversity. First, we rehearse some of the main types of argument relating to the practice of ecological restoration. This is followed by a description of the case study, and by a summary of what we take to be the main positions in the ongoing debate over reintroduction of beavers. We then interpret these different positions, asking in each case how "biodiversity" is being understood. In this way, we try to establish the causes of the disagreement. It is important to distinguish between disagreements caused by conflicting interests and disagreements caused by conflicting values. We shall focus on a special type of disagreement where there seems to be a genuine conflict of values pertaining to biodiversity. Finally, we show how the claim that biodiversity should be protected is made by several participants in the debate and taken to have remarkably different implications: the need to protect biodiversity has been invoked both in attacks on, and defenses of, reintroduction and other forms of ecological restoration.

11.2   THREE ATTITUDES TO ECOLOGICAL RESTORATION

One of the first modern and comprehensive definitions of ecological restoration was given by the Society for Ecological Restoration: "The intentional alteration of a site to establish a defined indigenous, historic ecosystem. The goal of the process is to emulate the structure, functioning, diversity and dynamism of the specified ecosystem" (Aronson et al. 1993).

Species reintroduction can be seen as a limited type of ecological restoration – a type used where a particular species is missing. According to guidelines developed by the World Conservation Union Re-introduction Specialist Group, reintroduction is an "attempt to establish a species in an area which was once part of its historical range, but from which it has been extirpated or become extinct" (IUCN 1995).[1] The overall aim of reintroduction is to establish viable, free-ranging populations in the wild of species that have become globally, or locally, extinct in the wild, and to do so with minimal commitment to long-term management. The term reestablishment is according to IUCN (1995) a synonym, but implies that the reintroduction has been successful. Sometimes, distinctions are drawn between restoration, rehabilitation, and reclamation. Definitions of these terms vary, and the differences between them are often not entirely clear. Rehabilitation may be defined as encompassing "a range of options which do not aim at exact fidelity to a predisturbance system" (Throop 2000, 13). However, the functioning and species composition of a rehabilitated system may be similar to the way they once were. Reclamation, on the other hand, is a process of conversion involving radical shifts in the structure of a system.

A more recent definition of ecological restoration, adopted by the Society for Ecological Restoration in 1996, reflects a shift in the goal of restoration from establishing a historically defined ecosystem to recovering ecological integrity: "Ecological restoration is the process of assisting the recovery and management of ecological integrity. Ecological integrity includes a critical range of variability in biodiversity, ecological processes and structures, regional and historical context, and sustainable cultural practices."[2] This more process-oriented goal undermines some of the criticisms that have been leveled at the previous definition. Instead of placing value specifically on the recovery of "natural balance," or on the recreation of a predisturbed state, the emphasis is, perhaps more modestly, on the repair of past damage. It has been claimed that, understood in this way, ecological restoration cannot be used as readily as an argument to justify current or forthcoming degradation (Cowell 1993).

These differing conceptions of ecological restoration have at times stirred up a rather harsh debate, especially among environmental philosophers (Woolley and McGinnis 2000; see also Mannison 1984; Elliot 1984; Katz 1991; Gunn 1991; Elliot 1994; and Katz 1996). Some negative views of the so-called restoration thesis are recapitulated by Elliot and Katz. The restoration thesis is the claim that any loss in the value of an area is only temporary and can in principle be compensated for later by the recreation of something of equal value. Elliot (1982) rejects this thesis and, using an analogy from the art world, describes restored areas as "fakes." One of his main claims is that naturalness cannot be restored if "natural" is defined as unmodified by human activity. According to Elliot (1997), an ecosystem's value is dependent upon its history – its having evolved out of natural processes.

Katz (1992), while accepting Elliot's main view, discusses some of the limitations in the art analogy. One of his claims is that the restorationist's use of the terminology of "repairing" ecosystems presupposes anthropocentrism and involves a fondness for technological fixes. According to Katz, restoration is part, not of the solution, but of the problem of continuing human domination. Katz (2000) argues that the human intentionality is what creates the distinction between human artifacts (e.g., restored ecosystems) and natural entities. We should understand "that there is a realm of value with which we should not interfere . . . We cannot be the masters of nature, molding nature to our wishes and desires, without destroying the value of nature" (38). (Cf. Birnbacher and Lee in this volume.)

According to Light (2000), however, a more productive response to the problem of restoration is to distinguish between so-called benevolent and malicious restorations. From this more pragmatic perspective, Light argues that Elliot's case focuses on malicious restoration. Such restoration acts in effect as an excuse for the deliberate damage of the natural environment. But benevolent restoration need not be a sign of human domination, as Katz has claimed. Instead, it may signify an intention to heal the relationship between human beings and nature. Moreover, Attfield (1994) asserts that our role in relation to nature is a dual one. First, we must act as preservers and restorers, because the full value of a predisturbed system can be recovered, provided that an array of former species can flourish in accordance with their nature. Second, our flourishing is important as well, and it is not necessarily a sign of domination. Rolston (1994) also supports the idea of restoration as part of a relationship with nature where intervention is inevitable. He claims, in contrast with Elliot, that ecological restoration can help to salvage values, and that natural values and naturalness do return. However, he concedes that for obvious reasons historical continuity cannot be recovered. Another important point is that many ecological restoration projects do not in fact attempt to restore ecosystems that are natural in the sense implying that the systems are humanly undisturbed and spontaneous. They aim to restore ecosystems that are natural in a culturally dependent way.

Table 11.1. *Three Attitudes Toward the Introduction and Reintroduction of Species*

	Attitude		
	Wise-use	**Pragmatic**	**Respect for nature**
Accepts species introduction	Yes	No	No
Accepts species reintroduction	Yes	Yes	No

In order to clarify the case study, we will distinguish between three standard attitudes to reintroduction (see Table 11.1). The first, which we call the *wise-use* attitude, has not been prominent in this particular debate. It is rooted in Pinchotian conservationism and represents an essentially anthropocentric ethical outlook, stressing the value of nature's use. According to this position, any species can in theory be introduced, or reintroduced, depending on its associated benefits and harms. First, the foreseeable negative consequences of a proposed introduction – for example, the damage done by the reintroduced species to forests and fields – should be determined. Second, perceived benefit of the introduction, that is, its use-value, should be assessed and balanced against the predictable negative consequences to decide whether introduction can be recommended. This attitude is the underlying rationale in game and fisheries management, where the anthropocentric commitment is evident and has justified the harvest of introduced species, as well as forest management and farm practices, throughout the last hundred years. Many of the present arguments for ecological restoration are in essence based on this attitude (cf. Throop 2000).

At the other end of the spectrum is an approach that might be named the *respect for nature* attitude. In this approach species introduction is opposed a priori.[3] Proponents of respect for nature look upon the human interference involved in restoration as yet another sign of human domination of nature (cf. Katz 2000). Reintroduction breaks up the historic continuity of a specific habitat or landscape. Both the reintroduction and (more seriously) the introduction of species amount to meddling with nature, and neither can be morally justified.[4]

Third, a combination of the two previous attitudes, a *pragmatic* attitude, can be discerned. Pragmatists oppose species introduction. However, they accept reintroduction, partly on wise-use grounds. In effect, they apply a form of environmental impact assessment here. From the philosophical point of view, reintroductions may be of a malicious or benevolent kind (cf. Light 2000). On the other hand, pragmatists agree with those who demand respect for nature that species introductions are neither acceptable nor desirable, whatever benefits arise. Reintroduction is seen as an exception to otherwise standard nature conservation practices. This somewhat radical departure could, for example, be justified where it is difficult for the species in question to migrate naturally to the country. In the following discussion, we will consider a real case. We shall review some of the actual reasoning attending this case and relate this to the three attitudes we have identified. Conflicts of interest are rampant in questions of reintroduction. An example would be the conflict between the interests of sports hunters and those of fish farmers. But the focus in the case study is on value conflict. True value conflicts occur when, for example, an environmentalist acknowledges intrinsic value in nature and a natural resource manager conceives of nature as only having instrumental value. The recognition of these differences in underlying value assumptions can contribute to our understanding of crucial differences in opinion regarding species restoration. Another fundamental clash is illustrated by the way biodiversity is used as an argument. The wise-use and extreme respect for nature positions both use it, but with entirely different outcomes.

11.3 CASE STUDY: REINTRODUCTION OF THE EURASIAN BEAVER

The Eurasian beaver (*Castor fiber*) is a semi-aquatic herbivorous rodent with webbed hind feet and a characteristic broad, flat, scaly tail. It is well known for constructing dams, dens, and partially submerged lodges, and was once abundant in forest zones and wooded river valleys in Europe and Asia (Andersen 2001). In the course of the last millennium, beavers have died out in many European countries. In Denmark, where our case study is located, beavers died out probably more than two thousand years ago, in the Bronze Age (1,800 to 500 B.C.). With increasing cattle husbandry, the prime beaver habitats – the wild meadows along small streams – were lost through their use for grazing and hay harvest (Aaris-Sørensen 1998). Moreover, habitats were generally degraded or disappeared as a result of population increase and subsequent growth in agricultural activity. These required extensive clearing of natural woodlands. Excessive hunting also contributed to the decline (Fritzbøger 1998). Beavers disappeared in Italy and Britain in the sixteenth century and in Sweden and Finland in the second half of the nineteenth century (Nolet and Rosell 1998). There were only five small populations of seven hundred animals total in Europe at the beginning of the twentieth century.[5] Today, bans on hunting, the establishment of wildlife sanctuaries and, since the mid-1920s, species reintroduction, have boosted the Eurasian

beaver population to approximately 350,000 animals. Most European countries where the beaver was once native have now reintroduced animals
from the few surviving populations in Europe. (See, e.g., Nolet and Baveco
1996; MacDonald 1995; MacDonald et al. 1995; Halley 1995; Mammal
Society 1999.)

In Denmark, a number of more or less directly involved interest groups
have a stake in the beaver's reintroduction.[6] First, there is the Danish Ministry
of Environment and Energy. This ministry has supported the reintroduction
plan. Second, there is the National Forest and Nature Agency, a government
body responsible for drafting management plans, implementing these, and
organizing public consultation. Third, landowners, such as woodland owners
and farmers, are likely to be directly affected by any plans involving reintroduction. And finally, special interest groups and nature conservation bodies
represent the interests of those, among the public, who desire input on the
issues affecting the natural environment.

11.3.1 Reasons Offered in Favor of Reintroducing Beavers

The National Forest and Nature Agency is responsible for the beaver reintroduction scheme. It offers two major reasons why the beaver should be
reintroduced (Asbirk 1998). First, there is an international legal responsibility to consider reintroduction if the beaver is unlikely to be able to migrate
naturally to part of its former range. The Eurasian beaver has a fragmented
distribution across its potential range, and this is interpreted as a sign of non-
favorable conservation status. Second, several benefits appear to arise from
reintroduction. The beaver is considered a keystone species. Such a species
plays a vital role in an ecosystem, for example by maintaining the diversity of
the ecosystem (Gilpin 1996). Beavers and their activities are likely to render
recreational enjoyment of nature more colorful.

The legal responsibility arises from the Bern Convention. Article 11(2)
of this Convention on the Conservation of European Wildlife and Natural
Habitats stipulates that:

> Each Contracting Party undertakes: (a) to encourage the reintroduction of native
> species of wild flora and fauna when this would contribute to the conservation
> of an endangered species, provided that a study is first made in the light of
> experiences of other Contracting Parties to establish that such reintroduction
> would be effective and acceptable.

The beaver is listed in Appendix III of the Bern Convention, which means
that appropriate and necessary legislative and administrative measures should

be taken to ensure its protection. However, this does not necessarily entail restoration in countries where it has become extinct. But in view of the biodiversity goals enshrined in the EEC Council Directive on the Conservation of Natural Habitats and Wild Fauna and Flora, a case for species restoration can be made (EEC Council Directive 92/43/EEC of 21 May 1992). Restoration should be considered with regard to species listed in annexes II and IV – that is, where the conservation status is judged "not favorable" and strict protection is needed. Implementing the provisions of this directive, member states shall, according to Article 22(a):

> Study the desirability of re-introducing species in Annex IV that are native to their territory where this might contribute to their conservation, provided that an investigation, also taking into account experience in other Member States or elsewhere, has established that such re-introduction contributes effectively to re-establishing these species at a favourable conservation status and that it takes place only after proper consultation of the public concerned.

The status of the Eurasian beaver on the global IUCN red list is not endangered but "Low risk: near threatened" (Asbirk 1998, 15). According to the IUCN (1994) Red List categories, a taxon is Lower Risk when "it has been evaluated, but does not satisfy the criteria for any of the categories Critically Endangered, Endangered or Vulnerable." The subcategory, Near Threatened, includes taxa which "do not qualify for Conservation Dependent [another subcategory in Lower Risk] but which are close to qualifying for Vulnerable." A taxon is Vulnerable when it is "facing a very high risk of extinction in the wild in the medium-term future." Here, it is a matter of debate whether the best conservation strategy is to repopulate most of the natural range or to concentrate on certain key areas (Nolet and Rossell 1998). It is a question of spatial scale – a question of whether to reintroduce in each of the countries in which the beaver once lived. Reintroduction is deemed necessary because it is almost impossible for the beaver to migrate naturally to certain countries in which it is absent. Sea surrounds Denmark on three sides, and the only possibility of natural migration is from the south, via Germany. However, this might prove difficult, because all the waterways run East–West or West–East, and many man-made artifacts such as roads, towns, and dry cultivated land block the way (Asbirk 1998).

Aside from the legal reasons, a few moral arguments in favor of beaver reintroduction have been given. Most other European countries have already reintroduced the beaver during the past eighty years, and now, the government suggests, Denmark should follow suit. The Eurasian beaver is native to the country. According to the government's National Forest and Nature Agency, it has a "right" to live there (Klein 1999b, 5).

But not only does Denmark have a legal, and perhaps moral, obligation to consider reintroduction, several expected benefits are connected with the reintroduction of beavers. It is a well-documented empirical fact that beavers will foster variation and stability because they are a keystone species in wetland habitats (Nolet and Rossell 1998; Andersen 1999). One of the main arguments put forward by the National Forest and Nature Agency is that the beaver will help to create more dynamics in nature: "It is not the beaver as a species which is the deciding factor, but the beaver as one of the most powerful driving forces in the most characteristic, original Danish nature types" (Klein 1999a, 6, our translation).

Beavers create open areas in wet woodland and thus help to increase a diversity of light-dependent flora. Threatened insects and mushrooms dependent on dead wood (which is rarely found in modern hardwood plantations) benefit from their tree-felling activity. The beavers might also prove useful as a new, sought-after game species, since relatively large numbers of people hunt for sport today. And in a broader perspective, beavers are likely to generate a high-quality recreational experience of nature of the kind currently in demand by the public at large in many Western European countries: "The beaver is an interesting animal that it is exciting to experience in nature. The beaver is able to habituate to boat traffic and the outdoor-life of human beings, so there are good opportunities to see or find its tracks" (Asbirk 1998, 23, our translation). In a situation where true wilderness areas characterized by natural dynamics are hard to find, other ways of making it possible for the public to enjoy so-called "quality nature experiences" need to be considered. The reintroduction of beavers will help to create natural dynamics and thus wilderness-like areas.

The main justification for the artificial return of beavers may be summarized as the fulfillment of legal, and to some extent moral, responsibilities; the prospect of benefits such as increased variation in nature, and the possibility of improved recreational experience of nature. According to opinion polls, animal rights groups, nature conservation groups, and a substantial sector of the public at large want to "help" threatened animal species and add variation to nature (Klein 1999b). However, while many have this general attitude, some serious reservations about reintroduction are also discernible.

11.3.2 *Reasons Offered against Reintroducing Beavers*

Opposition to species restoration comes from several quarters. Some opponents, such as farmers and recreational fishermen, fear the environmental impacts of the beavers. Others, such as some nature conservation groups, believe that beavers will have too little impact on the landscape and call for solutions that could lead to more substantial ecological change. These groups do commend beaver reintroduction, but they think comprehensive reintroduction policies need to be thought through first. An independent government advisory council also finds that policies need to be thought through before initiating reintroduction (Naturrådet 1998). The council generally argues that species restoration breaks natural continuity. Let us take a closer look at these arguments.

Landowners – for example, those with farms adjacent to proposed release sites – worry that beavers will do direct or indirect damage to trees, or, by causing flooding, wreck cultivated fields and fish farms. Some woodland owners and farmers fear that beavers will change the general appearance of old cultural landscapes. Special interest groups, such as the sports anglers, are concerned that fishing will be disturbed, and oppose reintroduction of the beaver. Moreover, the sports anglers want the current population of beavers removed from the country (Thygesen 2003). Even hunters, who generally welcome new game species, point out that considerable regulation of population (not hunters) might be needed, since the beaver's main natural enemy, the wolf, is absent in most parts of Western Europe: "[W]e will not be the authorities' 'dustman' . . . we like to go hunting, but we will not be human scavengers . . . it is important that a new species gets the opportunity to act naturally" (Steinar 1998, 8, our translation).

Nature conservation groups assert that beaver reintroduction, even if the beaver is a keystone species capable of bringing variation into ecosystems, is too limited. It will not lead to a much-needed general habitat improvement, as the blocking of drainpipes on old woodlands might. These groups question the argument that, as an ecologically important species, the beaver will be a significant generator of habitat restoration.

At a conceptual level, some conservation groups have claimed that the reintroduction of beavers by artificial means will leave no room for natural dynamics. They interpret natural dynamics as dynamics without human interference. From this it follows that the resulting dynamics created by beavers that are artificially introduced cannot be regarded as natural. Implicitly, of course, the non-natural is regarded as less valuable here than the natural. The claim is that non-natural migration is meddling with nature, which is presently not called for. National Nature and Forest Agency biologists have countered that, on the contrary, it is not natural that the beavers can no longer be found in the wild (Asbirk 1999, personal communication). Regardless of the soundness of this viewpoint, a governmental advisory body, the Danish Nature Council, and some nature conservation groups have argued that, lacking a consistent policy, "random" species restoration will fail to deliver a "naturally" functioning ecosystem. Instead, a member of the Council argues, such restoration turns nature into an open zoo or theme park: "Some of us get a feeling that isn't real . . . when I see that beaver, I will think of the originator of the idea . . . if I come to the Silkeborg lake district and see a beaver swimming around, maybe even with a collar, then it is a zoo" (Stensgaard 1998a, 3, our translation). Thus, it is stressed that historic continuity is imperative for the appreciation of beavers, for the valuing of biodiversity, and for admiring nature in general. The independent advisory government council points to the fact that for the last thirty years, habitat improvements have formed the basis of Danish conservation practices. The Council denies that species reintroduction can be justified on the grounds that it is likely to be difficult for beavers to migrate naturally to Denmark. The fact that there is a theoretical, albeit slim, possibility that some beavers would overcome the obstacles is sufficient to show that reintroduction should be opposed.[7]

11.4   THE ONGOING DEBATE AND THE THREE ATTITUDES TO REINTRODUCTION

From a management perspective – that is, either the wise-use or the pragmatic attitude – the ecological value of the beaver is very important. Restocking an animal such as the Eurasian beaver will not only protect a flagship species, it is argued, but enhance threatened biodiversity within the habitat. The beaver is considered part of the "original" fauna. Its presence will, it is claimed, help to restore the ecological integrity of a natural ecosystem. This notion of an original habitat type depends on an underlying value assumption. As part of a restoration scheme of the Eurasian beaver and subsequent restoration of wetland ecosystems, the reintroduction of the beaver is believed to lead to a more original habitat involving a higher level of biodiversity. This habitat is believed to be typical of the region's natural environment, that is, the situation before human settlement and overhunting occurred.

By contrast, from a user standpoint direct and indirect use-values, such as recreational and aesthetic values, are emphasized. It is evident that here it is not solely the protection status of the Eurasian beaver which is decisive. The beavers are reintroduced to habitats that are hardly prime beaver habitats and are in need of substantial restoration. Human presence is seen as a constant, a condition to which the beavers will have to become accustomed. At the same time beavers are treated as means to satisfy the human need, or desire, for nature-based recreational experience.

From an environmental policy perspective, it is our obligations to the international community and future generations (described earlier) that matter. Arguments drawing on these factors differ from justifications of reintroduction that focus on a species' instrumental value to humans. They stress the cultural and historical value of the beaver as part of the native wildlife heritage of Europe. Moreover, many of the legal justifications rest on the assumption that beavers are granted existence-value. It is apparently this that explains why measures against threatened species should be pursued. The underlying argument seems to be that if part of nature is destroyed – in this case, if an animal species is exterminated as a result of human activity – restoration is required. This view is shared by a Danish environmental NGO called Nepenthes. A member of Nepenthes argues that restoration ecology, which admittedly differs from natural processes, can in fact help to alleviate a shared sense of moral guilt over the destruction and degradation of the natural environments: "We say, we want this and that! It is not self-created nature, but it is exciting anyway. I find it far more constructive to go out and do something, instead of sitting back being ashamed" (Stensgaard 1998b, 4, our translation).

A moral rationale for the restoration process would attach significance to the making good, or correction, of some injury – in this case, damage inflicted by us on natural ecosystems. However, it is not entirely clear who the beneficiaries of such correction are. Are they contemporary humans, or future generations, or the populations of animals and plants in the restored ecosystem?

The reintroduction of the beaver forces us to ask whether restoration of the entire species array from the period following the last ice age is called for as part of a biodiversity conservation scheme. Should wolves be reintroduced, notwithstanding the fact that, in many European countries, wolves were regarded as pests and culled less than a century ago because of the threat they posed to livestock? There is no comprehensive, clear policy on mammal and predator reintroduction and natural migration. However, when it comes to questions of reintroducing predators such as the wolf, concerns over potential harm to humans feature prominently. Likewise, the migration of wild boar to a country like Denmark, which has large exports of agricultural products, forces us to consider the risk of spreading disease to livestock animals.

The argument that restoration practices turn natural environments into zoos is expressive of the respect for nature attitude. It presupposes that the evaluation and appreciation of natural areas and the biodiversity they contain depend upon a minute knowledge of local history and ecological processes. This knowledge has been described as "knowledge that can be acquired through education and experience, just as one learns the history of art" (Katz 1991, 92; cf. Elliot 1982). Historic continuity is broken when species are restored, and in this way spontaneity and authenticity are lost, according to this view. Instead, natural restoration – natural in the sense that it occurs without human assistance – is opted for, even if it takes decades, or perhaps centuries, for the animal in question to migrate across national borders unassisted.

It is clear from this analysis that the arguments in favor of reintroducing beavers are not purely ecological, but have underlying value assumptions. The opponents of species restoration question these assumptions and insist that the reintroduction issue cannot be settled on the basis of the instrumental value of the beaver. The value of the biodiversity the beaver might support, and the value of the landscape the beaver might shape, have to be considered carefully.

11.5 VALUES AND NOTIONS OF BIODIVERSITY

This last claim prompts us to ask what is meant by biodiversity. It is evident from the preceding analysis of the beaver case that many types of value are at stake when species reintroduction is advocated or opposed. The values include use-values (e.g., relating to the beaver's pelt and hunting as such) and aesthetic values (e.g., relating to the "cute" appearance of the beaver). Moreover, the ecological value of the beaver as a keystone species, its less tangible existence-value as a species, and the possible attribution of intrinsic value stressing its right to live, are also occasionally invoked.

The question is: which value counts when we are discussing species reintroduction, or more generally ecological restoration, in relation to biodiversity preservation? Are the relevant values of a nonintrinsic kind only? Such values are commonly associated with traditional management of the natural environment and the attempt to balance (direct or indirect) benefits against costs. Or do we have to include values other than the nonintrinsic kind when deciding whether to restore? These differences in underlying value questions are reflected in different notions of biodiversity.

Ecological restoration, including species restoration, is a tool to conserve biological diversity. Its advocates appeal to a notion of biodiversity in which species richness is stressed. The conservation goal here seems to be twofold, as the case with beaver reintroduction illustrates. One goal is the conservation of the beavers as a species. This assumes that establishing beavers in their entire former range will improve their long-term conservation status. The second objective concerns the conservation of the various threatened species that depend on the variation in wetland habitats which beavers are able to create and maintain. In this second objective, the value of the beavers is instrumental and dependent on the improvement of biodiversity.

The ultimate value of biodiversity is also instrumental, however, for biodiversity is valued as a means of improving the ecosystem's integrity, stability, and resiliency. It is therefore questionable whether it is imperative that a former native species perform these tasks. If the important factor is the role a species plays in maintaining biodiversity, there should be no problem in introducing some other species, provided the ecological role is the same. In relation to these issues, the wise-use and pragmatic attitudes appear in effect to take the same view.

In addition to an ecological counterargument stressing the need for a keystone species, another objection can be raised against this suggestion. Opponents of ecological restoration emphasize that the conservation of biodiversity is important only where it relates to the maintenance of natural processes. It is only where authenticity is preserved that the respect for nature attitude views biodiversity as valuable in itself. Thus, natural processes have to be retained as the basis for biodiversity, and historic continuity with the past must be upheld.

In effect, the advocates and antagonists of restoration invoke two quite different notions of biodiversity. One is linked to species richness and ecosystem integrity, and the other is connected with authenticity and natural processes. Advocates appeal to a notion of biodiversity stressing species richness, where the value of biodiversity is instrumental. Biodiversity is seen as a means of improving the integrity of the ecosystem. Opponents refer to a notion of biodiversity linked closely with the concept of authenticity. They emphasize the importance both of retaining natural processes as the basis for biodiversity and of maintaining natural continuity. These different notions of biodiversity influence the way in which a given ecosystem or species is valued, and indeed evaluated. A conceptual framework originally developed in political philosophy might prove useful in tackling questions about how best to understand the different notions of biodiversity.

11.5.1   *Valuing Biodiversity: 'End-State Principles' and 'Historical Principles'*

In his theory of distributive justice, the political philosopher Robert Nozick (1974) distinguishes what he calls *end-state principles* from *historical principles.* According to Nozick, a social situation is fair and just, judged by end-state principles, only if it involves a distribution of goods, which, irrespective of origin, displays a certain structure.[8] Thus, in order to assess whether a state of affairs concords with an end-state principle, we require no information about the way this state of affairs was brought about. On historical principles, by contrast, whether a state is legitimate depends on its historical evolution, or the way it was brought about. Here information on how the given state has arisen is not just relevant but essential to a determination of justice.

This distinction can be applied to the biodiversity issue. In wise-use and (to a certain extent) pragmatic approaches to reintroduction, end-state principles focusing on structure, stability, and functionality are used to determine the value of a specific ecosystem. A certain number and distribution of species will be indicative of the functionality, stability, and resiliency that is characteristic of the ecosystem. As long as this is secured, positive value can be assigned to the ecosystem and the biodiversity contained within it. In the respect for nature approach, on the other hand, end-state principles alone are insufficient to determine the value of an ecosystem, and historical principles have to be applied. Here, the value of the ecosystem depends on its history, how it came to be as it is.

Table 11.2 shows the relationship between these principles of evaluation and the differing conceptions of biodiversity presupposed in the wise-use and respect for nature approaches to restoration. According to restorationists, a wet woodland habitat with reintroduced beavers and other typical, but perhaps previously endangered, faunal and floral elements should be judged against a suitable, selected reference. A reference is here understood as an ecosystem exhibiting certain structural or functional elements believed to be representative of a "natural" ecosystem with minimal human intervention. Thus, the

Table 11.2. *Conceptions of the Nature and Value of Biodiversity and Principles of Evaluation: Their Relationship to Three Attitudes to Restoration*

	Wise-use and pragmatic attitude: restorationists	Respect for nature attitude: anti-restorationists
Conception of biodiversity	Species richness	Authenticity, natural processes
The value of biodiversity	Instrumental: adds to ecosystem integrity	Intrinsic
Principle of evaluation	End-state principle: ecosystem's stability, structure, and functionality	Historical principle: ecosystem's history and evolution

habitat may be judged favorably, regardless of any breaks in historical continuity, following restoration efforts. Anti-restorationists disagree with this. Facts about how the beavers actually came to be at the site would seriously affect their evaluation of the habitat. If the beavers were introduced, or reintroduced, that habitat would not possess the same value as it would have done, had the beavers migrated to the area without direct human interference. The beavers would presumably still add to the stability and resiliency of the ecosystem in the long term. They would probably help to conserve biodiversity as well. But the historical fact that introduction, or reintroduction, once took place would be a critical difference for them. Judged by historical principles the ecosystem would be, if not worthless, then at least less valuable than an authentic ecosystem.

However, the appeal to authenticity in cases of reintroduction in domesticated environments causes problems. If domesticated environments relate to wilderness areas in the same way as domesticated animals relate to wildlife, is an authentic dog best represented by a wolf? Clearly, it is difficult to decide where the demarcation line should be drawn. There is a long tradition of heavy utilization and manipulation, for example, drainage in many postindustrial societies. Since it is difficult, and in a European context in many cases futile, to restore an ecosystem to an early pristine or presettlement condition, ecological restoration is at best an exercise in approximation (Cairns 1995; see also Hobson and Bultitude in this volume).

Moreover, many species that are now considered native were introduced just a few centuries ago (Agger and Sandøe 1998). The current paradigm in ecology has replaced the idea of a "balance of nature" with an idea of a "flux in nature," and this too makes it difficult to identify authenticity (Pickett and Parker 1994; Aronson et al. 1995). The distinction between nature and culture has also become obscure, which again renders the notion that a habitat is authentic, or natural, or original problematic.[9] In many cases the best option seems to be to repair damage, or return an area to a former condition, and to acknowledge that this condition, being historically defined, is to some degree arbitrary.

11.6 CONCLUSIONS

At first glance, species reintroduction addresses biological, technical, and managerial issues. Beaver reintroduction has been justified primarily on an ecological basis: that is to say, it is defended on the grounds that it restores species richness and maintains evolutionary and ecological processes. However, as this case study illustrates, basic ethical questions regarding the origin and character of nature's value bear upon these factual issues. Appeals to the powerful concept of biodiversity are made by both the advocates and opponents of restoration, but as we have seen, with significantly different results.

In our view restoration practices can be as acceptable, and in many cases as necessary, as preservation efforts or wise-use policies involving minimal intervention, say, to protect certain species. However, the main lesson from the beaver case concerns the values underlying debates about restoration. Greater awareness of these values, and their promotion, is required. We suggest that a careful examination of the conflicting notions of biodiversity invoked in discussions of restoration policy and management will prove helpful in deciding whether, where, and what to restore.

NOTES

1. It may also concern a lower taxonomic unit, for example subspecies, if that can be unambiguously defined.
2. http://www.ser.org/definitions.html.
3. We do not use the phrase "respect for nature" in the way Taylor does (1986). Given respect for nature in Taylor's sense, certain principles of distributive and restitutive justice could permit reintroduction.
4. It important to note that those who almost never expect environmental and socio-economic costs to be met by sufficient benefits share views on species introduction and reintroduction, but for entirely different reasons. Conservative farmers and urbanites exemplify this NIMBY (Not In My BackYard) attitude.
5. Beaver hunting provided pelts, meat, and chemical substances derived from its castor sacs that were used both for medicine and as a base aroma in perfume.
6. The case study is based on, among other things, drafted reintroduction policies, proposed management plans, and statements from special interest groups, lot owners, and a governmental advisory council. The case is mainly based on the Danish process, but similar types of arguments can be found in discussions of species reintroduction in other countries in Europe.
7. Unlike in, for example, the United Kingdom, where it is most unlikely that beavers will arrive by means of natural migration (MacDonald 1995).
8. Utilitarian accounts of justice make use of an end-state principle: the classical formulation treats a distribution as just if it maximizes the overall quantum of happiness. Nozick's own theory of just acquisition deploys historical principles.
9. Cf. Light (2000), who talks about the "culture of nature."

BIBLIOGRAPHY

Aaris-Sørensen, K. 1998. *Danmarks forhistoriske dyreverden.* 3. udgave. København: Gyldendal.

Agger, P. and P. Sandøe. 1998. Use of "Red Lists" as an Indicator of Biodiversity. In *Cross-cultural Protection of Nature and the Environment*, F. Arler, J. Jensen, and I. Svennevig, eds. Odense: Odense University Press.

Andersen, I. F. 1999. Bæveren i Danmark. *Naturens Verden* 82(2):2–13.

2001. Bæveren. *Natur og Museum* 40(1):1–35.

Aronson, J., S. Dhillon, and E. Le Floc'h. 1995. On the Need to Select an Ecosystem of Reference, However Imperfect: A Reply to Pickett and Parker. *Restoration Ecology* 3(1):1–3.

Aronson, J., C. Floret, E. Le Floc'h, C. Ovalle, and R. Pontanier. 1993. Restoration and Rehabilitation of Degraded Ecosystems in Arid and Semi-Arid Lands. *I*. A View from the South. *Restoration Ecology* 1: 8–17.

Asbirk, S., ed. 1998. *Management Plan for the European Beaver* (Castor fiber) *in Denmark*. Copenhagen: The National Forest and Nature Agency.

Attfield, R. 1994. Rehabilitating Nature and Making Nature Habitable. In *Philosophy and the Natural Environment*, R. Atffield and A. Belsey, eds. Cambridge: Cambridge University Press. [Reprinted in W. Throop, 2000.]

Cairns, J. 1995. Restoration Ecology: Protecting Our National and Global Life Support Systems. In *Rehabilitating Damaged Ecosystems*, 2d ed., J. Cairns, ed. Boca Raton, FL: Lewis.

Cowell, C. M. 1993. Ecological Restoration and Environmental Ethics. *Environmental Ethics* 15:19–32.

Elliot, R. 1982. Faking Nature. *Inquiry* 25:81–93.

1984. The Value of Wild Nature. *Inquiry* 27:359–61.

1994. Extinction, Restoration, Naturalness. *Environmental Ethics* 16:135–144.

1997. *Faking Nature. The Ethics of Environmental Restoration*. London and New York: Routledge.

Fritzbøger, B. 1998. *Det åbne lands kulturhistorie ca. 1680–1980*. Frederiksberg: DSR-forlag.

Gilpin, A. 1996. *Dictionary of Environment and Sustainable Development*. New York: John Wiley & Sons.

Gunn, A. 1991. The Restoration of Species and Natural Environments. *Environmental Ethics* 13:291–310.

Halley, D. 1995. The Proposed Reintroduction of the Beaver to Britain. *Reintroduction News* 10:17–18.

Hobbs, R. J. and D. A. Norton. 1996. Towards a Conceptual Framework for Restoration Ecology. *Restoration Ecology* 4:93–110.

IUCN (The World Conservation Union). 1994. http://www.wcmc.org.uk/species/animals/categories.html.

1995. *IUCN Guidelines for Re-introductions*. Gland: IUCN.

Jordan, W. R. III. 1994. "Sunflower Forest": Ecological Restoration as the Basis for a New Environmental Paradigm. In *Beyond Preservation. Restoring and Inventing Landscapes*, A. D. Baldwin, J. de Luce, and C. Pletsch, eds. Minneapolis: University of Minnesota Press.

Kane, G. S. 1994. Restoration or Preservation? Reflections on a Clash of Environmental Philosophies. In *Beyond Preservation. Restoring and Inventing Landscapes*, A. D. Baldwin, J. de Luce, and C. Pletsch, eds. Minneapolis: University of Minnesota Press.

Katz, E. 1991. The Ethical Significance of Human Intervention in Nature. *Restoration & Management Notes* 9:90–6.

1992. The Big Lie: Human Restoration of Nature. *Research in Philosophy and Technology* 12:231–41.

1996. The Problem of Ecological Restoration. *Environmental Ethics* 18:222–4.

2000. Another Look at Restoration: Technology and Artificial Nature. In *Restoring Nature: Perspectives from the Social Sciences and Humanities*, P. H. Gobster and R. B. Hull, eds. Washington, DC: Island Press.

Klein, T. 1999a. Hvad er det med den bæver? *Weekendavisen* 29. januar – 4. februar 1999:6.

1999b. Bæveren gnaver sig ind på den danske natur. *Megafonen* Maj 1999:5.

Light, A. 2000. Ecological Restoration and the Culture of Nature: A Pragmatic Perspective. In *Restoring Nature: Perspectives from the Social Sciences and Humanities*, P. H. Gobster and R. B. Hull, eds. Washington DC: Island Press.

MacDonald, D. 1995. Beaver Come Home. *BBC Wildlife* Nov. 1995:72–6.

MacDonald, D., F. H. Tattersall, E. D. Brown, and D. Balharry. 1995. Reintroducing the European Beaver to Britain: Nostalgic Meddling or Restoring Biodiversity? *Mammal Review* 25:161–200.

Mammal Society 1999. Position Statement on the Re-introduction of the European Beaver, *Castor fiber*, into Scotland. Mammal Society: http://www.abdn.ac.uk/mammal/beavpos.htm.

Mannison, D. 1984. Nature May Be of No Value: A Reply to Elliot. *Inquiry* 27:233–5.

Nash, R. F. 1989. *The Rights of Nature. A History of Environmental Ethics*. Madison: University of Wisconsin Press.

Naturrådet 1998. Naturrådets syn på genindførelse af bæver i Danmark. SNS's debatmøde på Skovskolen, Nødebo 12. juni 1998. Naturrådet. Unpublished.

Nolet, B. A. and J. M. Baveco. 1996. Development and Viability of a Translocated Beaver Castor fiber Population in the Netherlands. *Biological Conservation* 75:125–37.

Nolet, B. A. and F. Rosell. 1998. Comeback of the Beaver Castor fiber: An Overview of Old and New Conservation Problems. *Biological Conservation* 5:165–73.

Nozick, R. 1974. *Anarchy, State, and Utopia*. Oxford: Blackwell.

OECD (Organisation for Economic Co-operation and Development). 1999. *Environmental Performance Reviews. Denmark*. Paris: OCED.

Pickett, S. T. A. and V. T. Parker. 1994. Avoiding the Old Pitfalls: Opportunities in a New Discipline. *Restoration Ecology* 2:75–9.

Rolston, H. 1994. *Conserving Natural Value*. New York: Columbia University Press.

Sheail, J., J. R. Treweek, and J. O. Mountford. 1997. The UK Transition from Nature Preservation to 'Creative Conservation.' *Environmental Conservation* 24:224–35.

Steinar, M. 1998. Bævere i Danmark? *Jæger* 7(2):6–8.

Stensgaard, P. 1998a. Vi laver lige noget natur. *Weekendavisen* 24.–30. juli 1998:3.

1998b. En skov i forfald er en god skov. *Weekendavisen* 31. juli–6. august 1998:4.

Taylor, P. W. 1986. *Respect for Nature: A Theory of Environmental Ethics*. Princeton, NJ: Princeton University Press.

Thomas, J. M. 1992. Restoration and the Western Tradition. *Restoration & Management Notes* 10:169–76.

Throop, W. 1997. The Rationale for Environmental Restoration. In *The Ecological Community*, R. Gottlieb, ed. New York and London: Routledge.

Throop, W., ed. 2000. *Environmental Restoration. Ethics, Theory, and Practice.* New York: Humanity Books.

Thygesen, J. K. 2003. Bæversagen skal afgøres i Vestre Landsret i 2003. Danmarks Sportsfiskerforbund. http://www.sportsfiskeren.dk/nyheder/2002/November/Baeversagen_skal_afgoeres_i_Vestre_Landsret_i_2003/

Tilman, D. 2000. Causes, Consequences and Ethics of Biodiversity. *Nature* 405:208–11.

Woolley, J. T. and M. V. McGinnis. 2000. The Conflicting Discourses of Restoration. *Society & Natural Resources* 13:339–57.

12

Differentiated Responsibilities

ROBIN ATTFIELD

12.1   INTRODUCTION

Many environmental thinkers maintain that environmental theory, including theories of environmental ethics, and political power have one thing in common: they need to be disaggregated and decentralized. The craving for generality, so disparaged by Wittgensteinian philosophers, becomes a punch-bag all over again for many defenders of biodiversity, whether pluralists, situationists, or postmodernists. "Show me a principle," they effectively say, "and I will show you an exception." While this sounds like a generalization itself, many apply it undaunted, for example, to purported principles of environmental obligation, stressing that environmental studies are characteristically if not essentially contextual, and are nothing without sensitivity to situations.

A praiseworthy example of this is provided by David Schmidtz's well-argued essay in *Environmental Values*, "Why Preservationism Doesn't Preserve." Schmidtz's essay belabors conservationist (or wise-use) principles as well as targeting preservationism in the way that the title leads the reader to expect, and shows how easily the pure pursuit of principle can in both cases undermine the environmental purist's objectives (Schmidtz 1997). The phenomenon is certainly widely recognized of principled people unintentionally undermining their own objectives, whether as efficiency experts, as librarians, or as parents; in the field of biodiversity preservation, a further example might be found in the refusal to countenance the extinction of any species anywhere, whatever the costs to humanity, including its poorer members, the stance that Wilfred Beckerman (1994) has labeled "Strong Sustainability," and Herman Daly (1995), "Absurdly Strong Sustainability."[1]

Thus, perhaps sensitivity to context, such as the contexts comprising "hot spots" of natural biodiversity, mostly situated in the Third World as they are, should supersede adherence to principle in matters of biodiversity preservation, granted that biodiversity is so unevenly distributed across the Earth. However, most protests against being guided by principle are nowhere near as persuasive as that of Schmidtz. In this chapter I discuss some cogent criticisms of universal principles, and consider to what extent sensitivity to context can be married to principled universalism in environmental contexts, and biodiversity contexts in particular. I shall argue that the Rio Declaration on Environment and Development makes some important contributions to resolving issues of preservation and of the responsibilities of different agents, groups, and countries.

As befits a discussion of across-the-board principles, I am using biodiversity in a broad sense, covering biological diversity of species, subspecies, and habitats alike, and including anthropogenic biodiversity (of landraces and cultivars, though not the products of genetic engineering) as well as natural biodiversity. The issues for this chapter are unaltered whether the units of biodiversity are species, subspecies, habitats, or landraces, and whether biodiversity itself is regarded as a state of affairs or a process. As I have argued in the past, diversity is not valuable intrinsically (Attfield 1991, 149–50), any more than individual organisms are,[2] or indeed the generality of their properties,[3] as opposed to their flourishing, including activities in which their essential capacities are developed.[4] However, there are, as I have argued more recently elsewhere (Attfield 1999a, chapter 8), ample grounds of a derivative kind for preserving biological diversity (in the sense used here). Hence, it is important here (as in other areas of applied ethics) to discover whether preservation involves following principles or discarding them so as to take full account of particularities, or being guided by both particularities and principles in some defensible combination.

12.2 CRITIQUES OF UNDIFFERENTIATED PRINCIPLES

Some kinds of theory lay themselves particularly open to charges of overgeneralization and of abstraction. Examples include theories that seek to generalize about the obligations of agents from hypothetical situations in which agents are described as having no ideals or life-plans, no family or friends, nor even any sense of their own identity, yet are equipped at the same time with as much knowledge as anyone could wish about the behavior of human beings and societies in general. Such are the agents of John Rawls's *A Theory of Justice* (1971), and it is widely claimed that little or nothing can be learned from the decisions that such agents might make for the decisions that real-life situated agents should make. Cogent reasons exist for Rawls's abstracting moves, for if the people of his original position had ideals, friends, or group identities, their selection of social principles might be made with partiality, favoring the prospects and (maybe) the environment of their own homeland or kindred. Yet the fact remains that once these abstractions have been made, they are scarcely impartial people, or even people at all, and that inductions from the supposed decisions of what Seyla Benhabib (1992) calls "disembedded and disembodied" people to those of embedded and embodied people may well be worthless. To cite a relevant example from Rawls, the Just Savings Principle, which involves passing down resources, including biological resources, to the next generation (Rawls 1971, section 44) undergoes no interpretative refinements with respect to issues such as whether the agents to whom it applies have the means to comply with this Principle without external assistance. Hence, even if legislators of moral principles would agree to this in the original position, its applicability to many (perhaps most) actual agents is questionable.

Overgeneralized theories about the obligations of each generation or every country are rightly open to charges of excessive abstraction. But suggestions that universal theories should be abandoned are also open to objection. Theories need to be context-sensitive without abandoning principle in favor of particularity. Benhabib, for example, does not despair of universal principles and universalistic theories. With a view to upholding some such principles and theories, and in the course of a variety of criticisms of those she rejects, she distinguishes two conceptions of self/other relations, the standpoint of the generalized other and the standpoint of the concrete other, and maintains that universalist moral theories can be rescued from overgeneralization and abstraction by (among other measures) avoiding the first conception (which she finds in Rawls, Kohlberg, and Habermas) and employing the second instead.

The first standpoint is depicted as follows:

> The standpoint of the generalized other requires us to view each and every individual as a rational being entitled to the same rights and duties we would want to ascribe to ourselves. In assuming the standpoint, we abstract from the individuality and concrete identity of the other. We assume that the other, like ourselves, is a being who has concrete needs, desires and affects, but that what constitutes his or her moral dignity is not what differentiates us from each other, but rather what we, as speaking and acting rational agents, have in common. (Benhabib 1992, 158–9)

On this basis, bundles of reciprocal but undifferentiated rights and obligations are prone to be elaborated. But all this contrasts with the second standpoint,

introduced as follows: The standpoint of the concrete other, by contrast, requires us to view each and every rational being as an individual with a concrete history, identity, and affective-emotional constitution. In assuming this
standpoint, we abstract from what constitutes our commonality, and focus on
individuality. We seek to comprehend the needs of the other, his or her moti-
vations, what that person searches for and desires. Our relation to the other is
governed by the norms of equity and complementary reciprocity: each is entitled to expect and to assume from the other forms of behavior through which
the other feels recognized and confirmed as a concrete, individual being with
specific needs, talents, and capacities (ibid., 159).

This standpoint recognizes not only other people's humanity but also their
human individuality. Individuals are understood as such not merely with regard to their capacity for agency, as if that were enough to individuate them
at all, but as people with a history, a concrete situation, concrete attachments,
and their own view of the good life. This standpoint supports an ethic in which
people are respected and their viewpoints are taken seriously; what Benhabib
has in mind is a version of Discourse Ethics, purified to remove Habermas's
tendency to return on occasion to the first standpoint (ibid., 169–70).

I am far from suggesting that Benhabib's solution to the problem of generalization be adopted wholesale. There are problems even with her exposition
of it. Thus, a standpoint that abstracts from "what constitutes our commonality" (including presumably from our being human) to "focus on individuality"
will have difficulty (if this is taken seriously) recognizing other people's humanity at all, and thus recognizing their human individuality as well. Besides,
her decision to focus on "each and every rational being" and then discover
the concreteness of each of them is virtually guaranteed to exclude nonhuman
creatures from moral consideration a priori, except where individual humans
happen to care about them. Relatedly, Discourse Ethics, with its focus on ideal
speech situations, is likely to exclude both voiceless creatures of the present
and the hitherto voiceless multitudes of the unborn of the future, both human and nonhuman. This approach certainly allows current agents to discuss
the treatment of such moral patients, but it fails to ensure that their interests
are taken seriously, or that they are represented. Indeed, unless representatives of voiceless moral patients are introduced into decision-making debates,
Discourse Ethics is likely to sell short both the future and other species as
well.

Nonetheless, the differentiating tendencies of Benhabib's preferred standpoint, that of the concrete other, offer hope of overcoming the overgeneralizing tendencies of universalist ethics, whether Kantian, rights-based, or
consequentialist. These differentiating tendencies have relevance not only

to expectations applied to individual agents, but to expectations applying to countries and governments as well. For example, principles of biodiversity preservation that take into account the differentiating features of diverse societies (different power, vulnerability, species concentrations, indebtedness, etc.) might prove viable, and a basis for international cooperation, where principles that ignore the differences would not. Thus, the problem of generality may arise not from universalism but rather from simplistic, overgeneralizing approaches, and its solution may lie in adopting principles that are sensitive to enough relevant differences, without ceasing to be universal.

Onora O'Neill, however, suggests that while idealized principles of justice are just as open to objection as the relativized principles of communitarians, there is nothing wrong with abstraction as such, for abstraction is indispensable for all reasoning about classes with common characteristics, as opposed to the detailed particularities of specific individuals. What is objectionable is idealized principles, ones which include characteristics that do not fit the cases that they are applied to. Rawls, for example, regards the individuals to whom his principles of justice apply as either heads of families or (at least) representatives of families, while families are assumed not to raise any issues of fairness or justice themselves. Principles based on assumptions such as these are to be regarded as idealized ones, and idealized principles apply only to idealized agents, whether sovereign individuals or sovereign states. (The same, I would suggest, applies also to idealized principles relating to generations, when a whole worldwide generation is represented as a single agent.) O'Neill believes, relatedly, that idealized conceptions of state sovereignty unduly limit discussions of international justice, often to the disadvantage of Third World women (O'Neill 1992, 52–61).

Instead, idealized principles are to be exposed and carefully avoided, and relevant differences of capacities and opportunities between agents given adequate acknowledgment (ibid., 67, 72). Rawls's difference principle and principle of equal opportunities, it might be suggested, seem to recognize these differences, and are certainly intended to do so. But they recognize them, I suggest, only for certain classes of moral patients, in the case of the difference principle those regarded by decision makers as worst off in whichever society (local or global) is being legislated for, while failing to ensure that all those capable of rational decision making are included among the decision makers of the original position, and respected accordingly. Besides, the heads of households engaged in the original position and its decision making may hold contestable beliefs about which members of society are the most disadvantaged and "worst off" in either local or world society, and thus about

241

which people stand to benefit from the very difference principle that is usually taken to ensure the minimizing of disadvantage for whomever are really worst off.

Further, whether or not O'Neill would recognize the following additional problem for Rawlsianism, no provision whatever is made, on standard interpretations of Rawls's theory, for the interests or well-being of nonhumans, despite their liability to be treated far worse than the most disadvantaged human beings. As for interpretations that suggest that the veil of ignorance should be regarded as including ignorance of one's species, such well-intentioned interpretations have little prospect of being recognized by the generality of Rawlsians, and they also suffer in acute form from the problems of invoking disembedded and disembodied legislators, likely to produce idealized principles. Here are legislators not only lacking any sense of belonging to a tradition or community but also unaware even of the species to which they belong, or thus of the nature that they will severally have when embodied. Their judgments about which basic goods to provide for in any human society, and in what order of priorities, are unlikely to command confidence or even to attain credibility.

However, as O'Neill insists, the rejection of idealized principles does not involve abandoning universal principles, without which disadvantaged people situated on the disadvantageous side (from their point of view) of political or cultural frontiers are liable to be abandoned in favor of relativizing endorsements of the status quo of one or another community or tradition (ibid., 60–2). Only on this basis will principles be found applicable to all the agents liable to interact with one another in the contemporary world (64–5), and in the absence of principles of this kind no reasons except ones of self-interest will be available to present to agents, whether individual or corporate, for playing their part in tasks that call for universal cooperation, such as that of preserving biodiversity.

Can the international environmental responsibilities of states in issues concerning biodiversity be understood on the differentiated but principled basis commended by O'Neill, and applied in a manner that respects the standpoint of the concrete other, as suggested by Benhabib? Certainly many theories of environmental obligations fail on these criteria, such as ones that generalize across generations, or across governments, without distinguishing between the circumstances of different countries and peoples; but I have discussed such matters elsewhere, and cannot discuss them further here (see Attfield 1999b; 1999a ch. 6, 10; 2003, ch. 6). In the next section I discuss instead a cluster of principles which appear to give these criteria international recognition already, since the Rio Declaration of 1992.

12.3 SOME DIFFERENTIATING PRINCIPLES

Here I introduce the seventh Principle from the Rio Declaration on Environment and Development:

> States shall cooperate in a spirit of global partnership to conserve, protect and restore the health and integrity of the earth's ecosystem. In view of the different contributions to global environmental degradation, states have common but *differentiated responsibilities*. The developed countries acknowledge the responsibility they bear in the international pursuit of sustainable development in view of the pressures their societies place on the global environment and of the technologies and financial resources they command. (Rio Declaration §7; Granberg-Michaelson 1992, 87)

What does this mean, and can it be sustained in the face of criticisms of "the craving for generality"? This is, after all, the agreed position of the governments of most countries on Earth, nominal as their agreement may in many cases have been. At the same time it grasps the thorny issue of the distribution of the burden of environmental responsibility, and it could be construed as implying that the burdens of different states and governments are far from unequal. In terms of global warming, for example, it could reasonably be taken to mean that while all states have some responsibilities, those that have emitted most greenhouse gases in the past or are emitting at the highest per capita rates in the present have a special responsibility to attain proportionate and sustainable levels across the future. And in terms of biodiversity preservation it might convey that states with limited economic resources but considerable biodiversity resources should be assisted by those with greater economic resources, particularly those whose policies or whose companies have contributed to biodiversity loss in recent decades.

Other Principles of the Rio Declaration can add to our understanding of Principle 7. Principle 6, for example, relates that:

> The special situation and needs of developing countries, particularly the least developed and those most environmentally vulnerable, shall be given special priority. International action in the field of environment and development should also address the interests and needs of all countries. (Rio Declaration §6; Granberg-Michaelson 1992, 87)

While the final sentence may seem to retract, in part, what is asserted in the previous sentence, the overall impression is that highly vulnerable countries with constrained capabilities should receive special assistance, albeit in ways that do not spell ruination for the others. Principles embodying these themes will simultaneously comply with O'Neill's concern to recognize differences of capacity and opportunity. They will also, as she further advocates, be ones which it would be possible for all parties to live by (O'Neill 1992, 64–5). Rather than weakening this stance, the final sentence is to be understood as stressing the importance in matters of environment and development of concerted international programs and policies, as opposed to ones limited by (often serendipitous and arbitrary) national boundaries.

In the four Principles that follow Principle 7, signatory states commit themselves to sustainable development and to the goal of eliminating unsustainable patterns of production and consumption, and to facilitate this through exchanges of science and of technology, and through subsidiarity, informed individual participation, and effective environmental legislation. Principle 11, the one about such legislation, proceeds to recognize that uniform policies and standards cannot be expected, in the following words:

> Environmental standards, management objectives and priorities shall reflect the environmental and development context to which they apply. Standards applied by some countries may be inappropriate and of unwarranted economic and social cost to other countries, in particular developing countries. (Rio Declaration §11; Granberg-Michaelson 1992, 87–8).

To some, this may sound like a concession that relativizes away all that has been achieved so far. Yet there is a sound general basis for at least the qualification about developing countries. One of the characteristics of underdeveloped countries is their current inability to provide for the basic human needs of their citizens or populations; but countries in such a situation are obligated, on most accounts of political obligation, to prioritize provision for these needs, and accordingly may well be obligated to assign a lower priority to biodiversity preservation. Where sustainable development is concerned, there could be argued to be little or no conflict between prioritizing human needs and doing so in a sustainable manner; but countries in this situation cannot be expected to prioritize preservationist policies that do not obviously cohere with such policies, particularly where countries have heavy debt-servicing obligations, unless resources of an unexpected magnitude become available from the Global Environmental Facility. Hence a qualification about the responsibilities of at least these countries needs to be entered if the resulting principles are to be remotely realistic, and to be applicable in their real-life situations.

Principle 15 introduces the Precautionary Principle in a similarly discriminating manner:

> In order to protect the environment, the precautionary approach shall be widely applied by states according to their capacities. Where there are threats of serious

or irreversible damage, lack of full scientific certainty shall not be used as a reason for postponing cost-effective measures to prevent environmental degradation. (Rio Declaration §15; Granberg-Michaelson 1992, 88)

While the requirement that the measures to prevent environmental degradation be cost-effective is regrettable in circumstances where there is as yet no scientific closure by which the cost of not taking these measures could itself be measured, this principle discriminates wisely in laying burdens of obligation only on those states capable of bearing them. It could also be held implicitly to mandate the more powerful states to act to prevent serious or irreversible environmental degradation abroad by companies subject to their jurisdiction, again in advance of scientific consensus, thus once again proportioning responsibility to power.

It should further be noted that the ensuing Principles 20, 21, and 22 draw attention to the roles in the pursuit of sustainable development of women, of young people, and of indigenous people and their communities. In the latter case, states are urged "to recognise and duly support their identity, culture and interests and enable their effective participation in the achievement of sustainable development" (Rio Declaration §22; Granberg-Michaelson 1992, 89). The presence of indigenous people and their communities among the citizens of a state's territories is, I suggest, rightly regarded as a general ground for such special obligations.

The conclusion seems to be justified that the much neglected Rio Declaration comes up with some defensible universal principles, sensitively tailored to differences of ability, vulnerability, power, and other general kinds of special circumstances. Further discussion, however, is in place before it can be concluded that they adopt the approach of the concrete other, and that they abstract without idealizing.

12.4 SATISFYING THE CRITICS

The standpoint of the concrete other contrasts with that of the generalized other, regarded as a rational individual (or sovereign state) "entitled to the same rights and duties we would want to ascribe to ourselves" (Benhabib 1992, 158). Within the standpoint of the generalized other, only what they and we have in common is taken into account. But the Principles of the Rio Declaration previously cited foreswear this standpoint through emphasis on "differentiated responsibilities" (Principle 7), "the special situation and needs of developing countries" (Principle 6), objectives and priorities reflecting "the environmental and development context to which they apply" (Principle 11), and the precautionary approach that is to be "applied by states according to their capacities" (Principle 15). Relevant differences are thus respected as well as commonalities, and thus to some degree the standpoint of the concrete other is adopted, with its concern to view each and every rational agent (or country) as an individual with a concrete history and identity. Whether enough regard is given to the national equivalent of an individual's affective-emotional constitution (another of Benhabib's requirements), which might be regarded as its culture or cultures, is less clear. But at least special attention is explicitly paid to recognizing the claims and the role of "the identity, culture and interests" of indigenous cultures (Principle 22).

For the same reasons, the principles cited are not "idealized principles" such as those censured by O'Neill, but ones wherein "relevant differences of capacities and opportunities between agents" are "given adequate acknowledgement" (O'Neill 1992, 72). So mindful are these principles of contexts as at least to raise the suggestion that their universal substance has been contextualized away, a possibility that I mentioned with Principle 6; but so clear is the emphasis on the rights and role of women (Principle 20) and of the young (Principle 21), on the precautionary approach (Principle 15) and, more subtly, on preserving the health and integrity of (what is called) the "earth's ecosystem" (presumably the biosphere) (Principle 7) that this accusation cannot be upheld. The charge that the Declaration is speciesist and thus wrongly grounded has greater plausibility, but can partly be met by reference to the aim of the same Principle ("to conserve, protect and restore the health and integrity of the earth's ecosystem"). Thus, the principles cited do seem to abstract without idealizing.

Abstraction can become objectionable when principles are introduced which fail to specify the different roles of individuals and communities as opposed to those of countries and of corporations. Fortunately the Principles go a long way to avoid this fault by stressing the distinctive roles of women, the young, and indigenous communities, as well as those of governments, although they could say much more than they do about corporations. The context of the precautionary approach is one where greater specificity might be in place about the different levels of responsibility involved, but the Declaration lays the task of reflecting on such matters on us all, and cannot be expected to resolve them at a stroke. (For further work on the precautionary principle, see O'Riordan and Cameron 1994; Attfield 2001).

Perhaps, though, abstracting is itself misguided, and an adequately localized world system would be better advised to eschew principles altogether. To this objection I have already given some of O'Neill's replies: in the absence of universal principles and of their recognition, vulnerable parties like

Third World women are liable to be oppressed or eroded. (In the present context it may be appropriate to say the same about vulnerable species that are not confined to one locality.) Further, principles are needed if guidance is to be available about how to proceed, whether locally or internationally. In their absence, too much is left to individual intuitions, which will often be massively divergent. Besides, even theories that stress decentralization and local loyalty rely, in part, on this being the best way forward quite generally for all the affected parties, if followed as widely as possible, and thus on a general principle. Thus, compliance with the requirements attributed above to O'Neill remains a matter for praise, and not blame (as critics of principles might suggest). For context-sensitive distinctions, such as those specified in the Principles cited, actually cut across or bestride contexts, drawing attention to the kinds of contexts that warrant special attention, and accordingly they are fully consistent with universal principles. Hence, as O'Neill stresses, we need not abandon principle in the cause of respect for particularity.

A more detailed argument than is supplied in the preceding remarks would be necessary to show how best to comply with Benhabib and O'Neill's requirements. It is nevertheless my hope that enough has been said to exhibit the possibility of satisfying them.

12.5 COOPERATION THROUGH DIFFERENTIATED RESPONSIBILITIES

Finally, I want to suggest that suitably sensitive principles embodying differentiated responsibilities (or at least some of them) could fit into a system of worldwide cooperation, as O'Neill clearly supposes, and as Principle 6 of the Rio Declaration (among others) exhorts. This conclusion warrants a more detailed argument than can be supplied here, but it is implicit in my earlier claim that principles of preservation that take into account the differentiating features of different societies might prove viable (that is, globally acceptable and workable), and might also prove to be a basis for international cooperation, where principles that ignore differences would not. Evidence that mutually beneficial cooperation can stabilize and spread both between individuals and between governments, however egoistic, even in the absence of mutual understanding or trust, has been supplied by Robert Axelrod in his intriguing work *The Evolution of Co-operation* (1984). Meanwhile, the credibility of my claims about principles embodying differentiated responsibilities is enhanced by the fact that some suitably sensitive such principles will concern burden-sharing for the common good, and that where a sufficient number and range of governments are prepared to endorse such principles, a basis already exists for a system of international cooperation.

All this can be illustrated briefly from the case of global warming. Granted the case for curtailing the growth of global greenhouse gas emissions and if possible for reducing such emissions, the Kyoto agreement of 1997 was based on the principle that the developed nations should reduce their emission levels of 1990 by (on average) 5.2 percent. In some ways this was a differentiating agreement; for example, Russia and the Ukraine were permitted not to reduce their 1990 emissions because of their special circumstances, and to sell the surplus quotas that the 1990 principle allowed them; and developing countries were exempted from the agreement altogether, at least in the first instance, for such reasons as that they need to increase their carbon emissions if they are to generate the electricity necessary to provide for the unsatisfied needs of their populations. On the other hand, the implicit principle that nations are entitled to continue polluting up to a high percentage of their 1990 emissions is in other respects an undiscriminating principle, entirely unsuitable as the basis of a global system, partly because its universal adoption would prevent nations other than the heavy polluters of 1990 from generating the electricity that development is likely to require in coming decades.

Some quite different principle is necessary as the basis of a system of worldwide cooperation. Mark Sagoff (1999) has recently suggested proportioning quotas to GDP; but this suggestion, though less arbitrary than the 1990 principle, takes even less account of the circumstances of the Third World than that principle, if globally applied, would do. But if, as Michael Grubb has suggested, the equal right of each person to access to the absorptive capacities of the atmosphere were recognized, and the principle were adopted of proportioning quotas to the population of each state, then a basis would exist which would take into account the special circumstances of populous developing countries, and which, through a system of quota-trading, would give developed countries an incentive to reduce pollution, while allowing them to purchase surplus quotas and simultaneously to subsidize Third World development (Grubb 1989, 1990). While this principle too has its problems, it appears to constitute an example of a principle that allows for relevant differences, it recognizes the concrete needs of each individual and of each state, and it could form the basis of a regime of international cooperation that might save the coastal plains and the small islands from inundation, together with their human and nonhuman inhabitants.

In the future, humanity and fellow-species alike will be dependent on such international treaties and regimes. Besides, recognition of human trusteeship of the planet involves (as I have argued in *The Ethics of the Global*

Environment 1999a) a readiness on the part of all countries to play their part. But playing their part does not involve uniform responsibilities (any more than the responsibilities of entire generations do), but responsibilities differentiated by the differing powers and resources of the various states and communities. Furthermore, without this kind of global cooperation, and without the application of such agreements at all political levels, national and local included, biodiversity is unlikely to be sufficiently preserved, human needs are unlikely to be sufficiently satisfied, and environmental justice is unlikely to be done.

NOTES

1. Stances which, like "Absurdly Strong Sustainability," prohibit substitution are criticized in Dieter Birnbacher's contribution to this volume.
2. Here I agree with Birnbacher (in this volume) that value-theories that attribute intrinsic value to individuals tend to be deontological theories in disguise; I have argued along parallel lines in "Intrinsic Value and Transgenic Animals" (Attfield 1998, 172–89). I also concur with Birnbacher's view that our relations with individuals do not make those individuals valuable intrinsically rather than extrinsically. But I find his view that such relations are themselves intrinsically valuable a puzzling one, which needs further justification.
3. Birnbacher's suggestion (in this volume) that it might be properties such as their beauty or their rarity that ground ascriptions of intrinsic value to individual creatures seems unconvincing, and it is unsurprising that he finds it problematic. More plausibly, what has intrinsic value is their flourishing, or attaining their own good (which is more a state of affairs than a property); and this state is in one sense unsubstitutable (because the good of one creature cannot be attained by any other creature), while it is certainly substitutable in another sense, in that one flourishing creature might be replaced by another flourishing creature without an overall loss of value, except where contingent losses are involved, and except where the flourishing of the first creature involves an autonomous desire to continue leading a worthwhile life.
4. Attfield 1995, chapters 4–6. This position would largely tally with the sentientism for which Kate Rawles (in this volume) expresses sympathy, without necessarily clashing with concern to preserve species and habitats, with which Rawles contrasts such sentientism.

BIBLIOGRAPHY

Attfield, R. 1991. *The Ethics of Environmental Concern*, 2d ed. Athens, GA and London: University of Georgia Press.

1995. *Value, Obligation and Meta-Ethics*. Amsterdam and Atlanta, GA: Éditions Rodopi.

1998. Intrinsic Value and Transgenic Animals. In *Animal Biotechnology and Ethics*, A. Johnson and A. Holland, eds. London: Chapman and Hall.

1999a. *The Ethics of the Global Environment*. Edinburgh: Edinburgh University Press and West Lafayette, IN: Purdue University Press.

1999b. Depth, Trusteeship and Redistribution. In *Proceedings of the XXth World Congress of Philosophy, Volume 1, Ethics*, K. Brinkmann, ed. Bowling Green, OH: Philosophy Documentation Center, 159–68.

2001. To Do No Harm: The Precautionary Principle and Moral Values. *Reason in Practice* 1.3:11–20.

2003. *Environmental Ethics. An Overview for the Twenty-first Century*. Cambridge: Polity Press.

Attfield, R. and B. Wilkins, eds. 1992. *International Justice and the Third World: Studies in the Philosophy of Development*. London and New York: Routledge.

Axelrod, R. 1984. *The Evolution of Co-operation*. New York: Basic Books.

Beckerman, W. 1994. Sustainable Development: Is It a Useful Concept? *Environmental Values* 3:191–204.

Benhabib, S. 1992. *Situating the Self: Gender, Community and Postmodernism in Contemporary Ethics*. New York: Routledge.

Daly, H. 1995. On Wilfred Beckerman's Critique of Sustainable Development. *Environmental Values* 4:49–55.

Granberg-Michaelson, W. 1992. *Redeeming the Creation*. Geneva: WCC Publications.

Grubb, M. 1989. *The Greenhouse Effect: Negotiating Targets*. London: Royal Institute of International Affairs.

1990. *Energy Policies and the Greenhouse Effect*. Aldershot: Gower.

O'Neill, O. 1992. Justice, Gender and International Boundaries. In *International Justice and the Third World: Studies in the Philosophy of Development*, R. Attfield and B. Wilkins, eds. London and New York: Routledge.

O'Riordan, T. and J. Cameron, eds. 1994. *Interpreting the Precautionary Principle*. London: Cameron and May.

Rawls, J. 1971. *A Theory of Justice*. Cambridge, MA: Harvard University Press.

Rio Declaration on Environment and Development, reprinted as an Appendix in W. Granberg-Michaelson, *Redeeming the Creation*, Geneva: WCC Publications, 1992.

Sagoff, M. 1999. Controlling Global Climate: The Debate over Pollution Trading. *Report from the Institute for Philosophy and Public Policy* 19.1:1–6.

Schmidtz, D. 1997. Why Preservationism Doesn't Preserve. *Environmental Values* 6:327–39.

Index

www.ingramcontent.com/pod-product-compliance
Ingram Content Group UK Ltd.
Pitfield, Milton Keynes, MK11 3LW, UK
UKHW040704180125
453697UK00010B/391